383 Repair and Servicing of Road Vehicles Series
Levels 2 & 3

Transmission, Chassis and Related Systems

AUTHOR **JOHN WHIPP**   EDITED BY **ROY BROOKS**

MACMILLAN

First published 1990

Published by
MACMILLAN EDUCATION LTD
Houndmills, Basingstoke, Hampshire RG21 2XS
and London
Companies and representatives
throughout the world

Printed in Great Britain by H. Shanley (Printers) Ltd., Bolton, Lancashire.

British Library Cataloguing in Publication Data
Whipp, J. (John)
383 Repair and servicing of road vehicles.
Levels 2 & 3, Transmission, chassis and related systems.
1. Motor Vehicles. Maintenance & Repair
I. Title II. Brooks, R. (Roy), 1929–
629.287
ISBN O–333–54297–5

Contents

Preface

This book, along with its companion volume, *Engines and Related Systems*, have been designed specifically to meet the needs of students following the City and Guilds 383 Syllabus, Repair and Servicing of Road Vehicles, at levels 2 and 3, and those taking other similar courses.

Undoubtedly the 383 syllabus is substantially different from its predecessors. Along with the very rapid advances in motor vehicle technology, it presents particular challenges by way of interpretation and implementation. There are also problems, with the syllabus, in avoiding overlap within the various elements, units and levels.

Much careful thought and considerable experience has gone into the content and style of these pages. There are, of course, areas where the editorial team would have preferred to have tackled certain topics somewhat differently, but were constrained by syllabus requirements. Equally, however, there are some instances where desirable additional material has been included as an aid to greater understanding.

Naturally enough these books are concerned, as were their highly successful predecessors, with the classroom aspects of motor vehicle work. They do, though, help the teacher to conveniently provide a level of understanding that will enable students to achieve a confident knowledge of systems and components, and their operation in the work situation. In consequence, workshop tasks are not covered in fine detail. It is expected that such procedures will be taught at work or in garage practice classes. Similarly it is also expected that students will refer to appropriate workshop manuals and/or instruction sheets for the work in hand.

By a combination of such methods and proper completion of these books, a very high standard of training will be achieved. This should be fully sufficient to satisfy examination requirements, as well as providing students with a valuable source of reference and a permanent record of their studies.

Each book is presented in a sound, logical order, with syllabus references shown for convenience. It should, though, be realised that this is not necessarily always the most appropriate teaching order, which is likely to vary considerably according to available facilities and preferences.

The editor and authors wish success to everyone who uses these books. If you have any constructive suggestions that you would care to make about them, we should be most pleased to hear from you via Macmillan Education Ltd, Basingstoke.

Roy Brooks
(Editor)

Acknowledgements

The editor, author and publishers would like to thank all who helped so generously with information, assistance, illustrations and inspiration. In particular the book's principal illustrator, Harvey Dearden (previously principal lecturer in Motor Vehicle Subjects, Moston College of Further Eduction); colleagues of the Burnley College and the North Manchester College; Andrea Whipp for her dedication in preparing the manuscript; and the persons, firms and organisations listed below. Should there be any omissions, they are completely unintentional.

A-C Delco Division of General Motors Ltd
Alfa Romeo (Great Britain) Ltd
Austin Rover Group
Automotive Products plc
Robert Bosch Ltd
R Boughton Esq.
British Standards Institution
Castrol UK Ltd
Champion Sparking Plug Co. Ltd
Citroën UK Ltd
City & Guilds of London Institute
Clayton Dewandre Ltd
Dunlop (SP Tyres UK Ltd)
Fiat Auto (UK) Ltd
Ford Motor Co. Ltd
Girling Ltd
Honda UK Ltd
Land Rover Ltd
Lucas Industries plc

Luminetion Ltd
MAN – Volkswagen
Metalistic Ltd
Michelin Tyre plc
Mitsubishi Motors (The Colt Car Co. Ltd)
Peugeot Talbot Motor Co. Ltd
Renault UK Ltd
Ripaults Ltd
Scania (Great Britain) Ltd
Seddon Atkinson Ltd
Suzuki Cars (GB) Ltd
Alfred Teves GmbH
Unipart Group of Companies
VAG (United Kingdom) Ltd
Vauxhall Motor Co. Ltd
Volvo Concessionaires Ltd
Westinghouse CVB Ltd
ZF Gears (GB) Ltd

Chapter 1

Clutches

| | ELEMENTS 19/36 | UNITS 16/35/36 | |

CLUTCHES

In a vehicle transmission system the clutch transmits the engine torque to the gearbox. State three other purposes which the clutch fulfils.

....
....
....
....
....
....

The clutch assemblies shown opposite are typical examples of light and heavy vehicle units, which are manually operated by the driver controlling a pedal. Label the illustrations and briefly describe the significant features of the clutch operation.

DRIVE

....
....
....

RELEASE

....
....
....

The main differences between clutch assemblies other than coil or diaphragm spring operation are:

diameter (increased for greater torque capacity)
clamping springs (increased in strength or number for increased torque capacity)
withdrawal mechanism (push or pull)
drive to pressure plate (lug or strap).

DIAPHRAGM SPRING TYPE

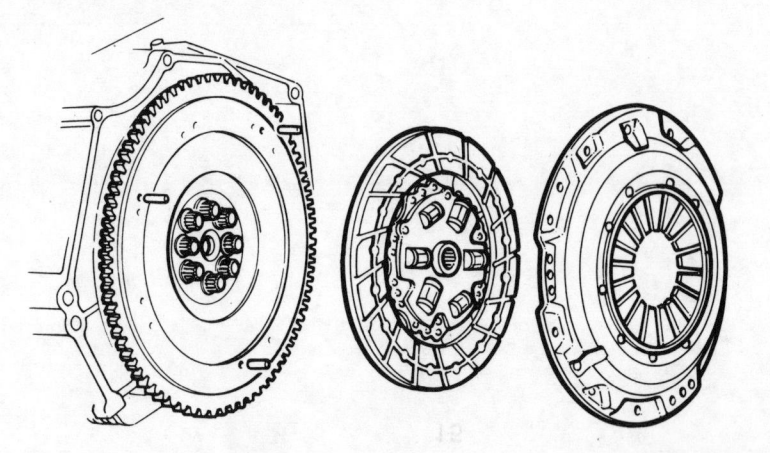

COIL SPRING TYPE (hgv)

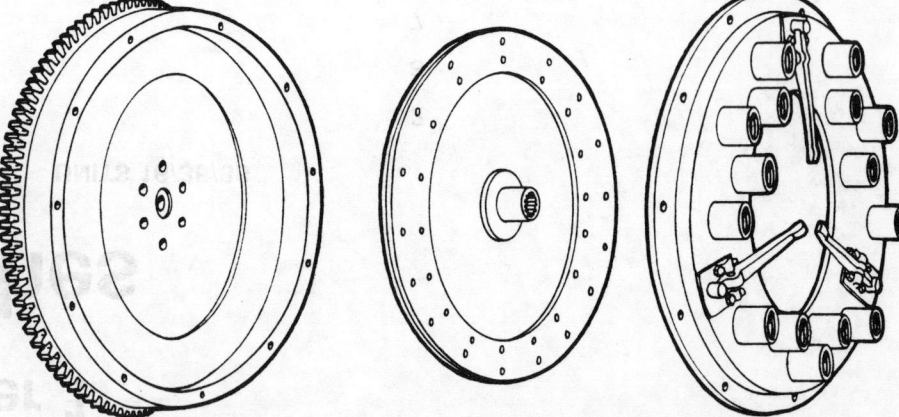

2

THE MULTI-SPRING CLUTCH

A simple multi-spring type of clutch is illustrated below. Label the drawing and state the function and operation of the parts listed opposite.

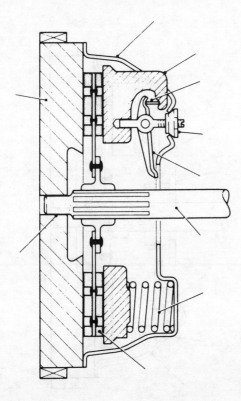

Power flow

In the clutch shown, 'lugs' formed on the pressure plate fit into rectangular slots in the cover to provide drive from cover to pressure plate.

Add arrows to the drawing to indicate the 'power flow' through the clutch from the flywheel to the primary shaft.

Cover

....
....
....

Coil springs

....
....
....

Pressure plate

....
....
....

Release levers – eye bolts and pins

....
....
....

Adjusting nuts

....
....
....

'Knife-edge' struts

....
....
....

19.1 THE DIAPHRAGM SPRING CLUTCH

A single-plate diaphragm clutch is shown below; label the drawing and add arrows to show the power flow through the clutch from the flywheel to the primary shaft.

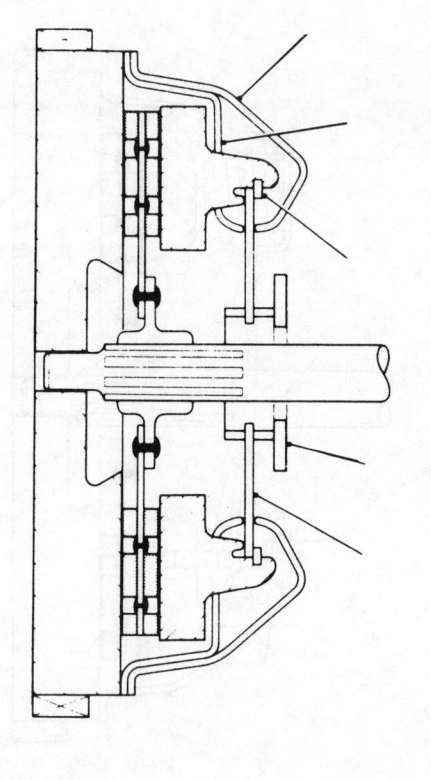

In the diaphragm spring clutch shown, state how the pressure plate is withdrawn during disengagement.

....

....

....

Spring characteristics

(Coil and diaphragm)

The graph below shows a typical load/displacement curve for a diaphragm spring. Show on the same graph, a typical load/displacement curve for a coil spring.

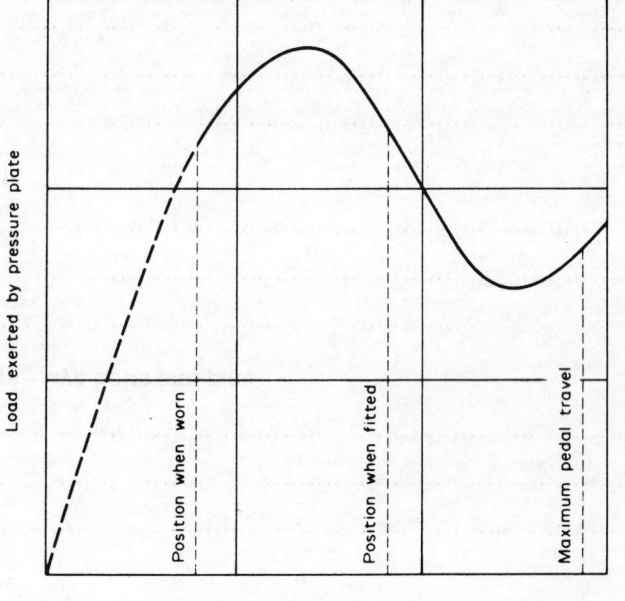

Load exerted by pressure plate — Position when worn — Position when fitted — Maximum pedal travel

Displacement of pressure plate

Study the completed graph and state two advantages of the diaphragm spring over the coil spring with regard to spring load in relation to pressure plate displacement.

1. ....

....

....

2. ....

....

4

STRAP-DRIVE PRESSURE PLATE

In the clutches shown so far the drive is transmitted via lugs formed on the pressure plate. Examine a 'strapdrive' clutch and make a simple sketch to illustrate the drive arrangement.

State the main advantage of the strap-drive system.

....

....

....

....

....

Why are flexible straps used to transmit the drive?

....

....

....

CLUTCH PLATE CONSTRUCTIONAL FEATURES

The two main types of clutch plate in general use are shown below.

(a)   (b)

Type (a) ....

....

Type (b) ....

Application

The spring centre type is used on most vehicles.
Give two examples of the use of the 'one-piece' clutch plate, stating why it is chosen for the particular applications.

....

....

....

....

....

....

....

Spring centre type

Label the component parts of the spring centre clutch hub shown below.

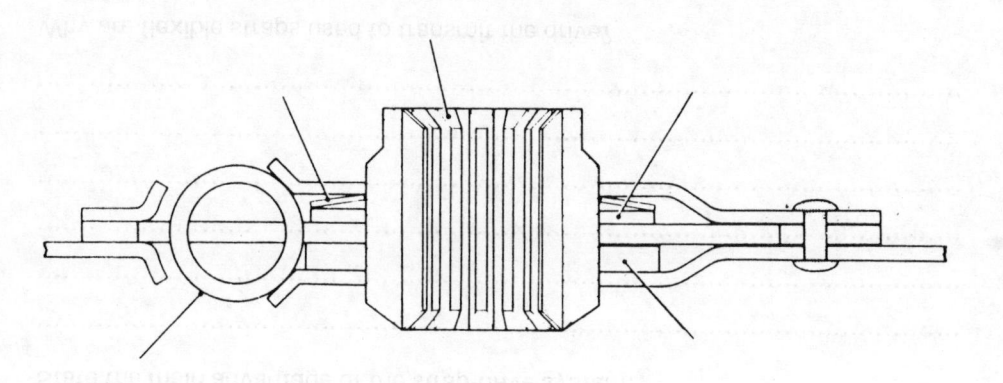

State the function of the various parts of the spring centre:

1. Coil springs: ....
....
....
....

2. Friction material: ....
....
....
....

3. Belville washer: ....
....

Friction materials – clutch plate facings or linings

The friction material from which the clutch linings are made must: maintain its frictional qualities at high temperatures; be hard wearing; withstand high pressure and centrifugal force; and must also offer some resistance to oil impregnation.

Several materials are currently in use for friction linings. Asbestos bonded with resin has been the most widely adopted material. Name two other ORGANIC (resin bonded) materials used for clutch plate facings.

....

An alternative to the organic type of lining is used in many heavy duty applications; this is ....

To help provide for smooth, progressive clutch engagement a certain amount of compressibility is built into the periphery of the clutch plate, where the clutch lining is secured to it.

Examine a clutch plate with some of the facing (that is, lining) removed and, with the aid of simple sketches, show in the space below how this compressibility is achieved.

CLUTCH OPERATION AND ADJUSTMENT

The clutch can be operated from the pedal through any of the following systems:

Mechanical (rods)
Mechanical (cable)
Hydraulic.

In many heavy goods vehicles the disengagement mechanism is compressed air assisted.

The clutch-release bearing is carried on a withdrawal lever or fork which pivots in the clutch housing. One arrangement is shown at (a) below; make a similar sketch at (b) to show an alternative.

Whichever system of operation is used an amount of free movement is necessary. State the reason for this.

....

....

....

....

Complete the drawing below to show how a clutch can be hydraulically operated; include provision for adjustment.

Clutch master cylinder

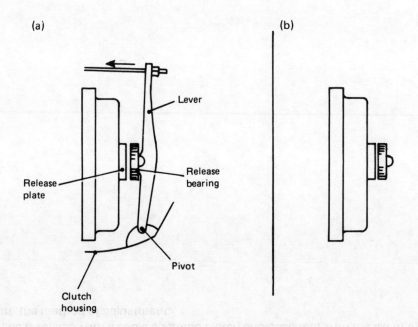

(a)   (b)

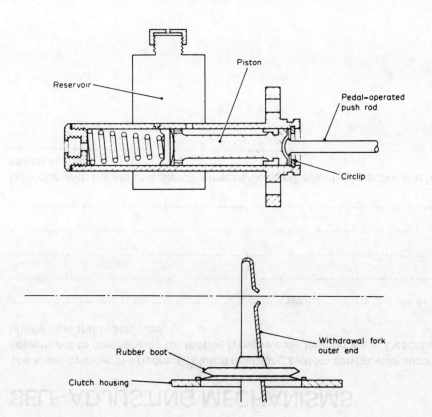

INVESTIGATION

Examine a vehicle with a cable operated clutch and describe, with the aid of a sketch, the method of adjustment.

....

....

....

....

....

State clutch faults which would create the need for:

(a) increasing free movement

....

(b) decreasing free movement

....

SELF-ADJUSTING MECHANISMS

The slave cylinder in a hydraulic clutch operating system can provide automatic adjustment to compensate for friction lining release bearing wear. Describe briefly how this is achieved.

....

....

....

....

....

Describe with the aid of a sketch a mechanical self-adjusting mechanism for a clutch system.

AIR ASSISTED CLUTCH OPERATION (hgv)

Power assisted clutch operation reduces the effort required at the pedal. One arrangement employs a clutch servo, which is an air/hydraulic unit. The servo is located at the clutch housing to actuate the clutch shaft lever. Study the section of a clutch servo below and describe its operation.

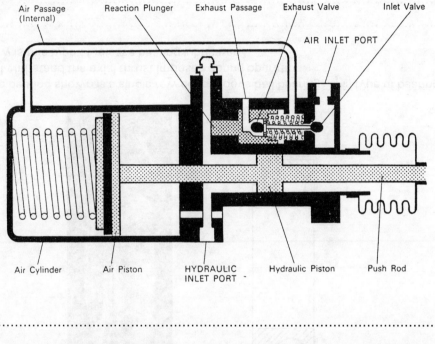

Air Passage (Internal) — Reaction Plunger — Exhaust Passage — Exhaust Valve — Inlet Valve

AIR INLET PORT

Air Cylinder — Air Piston — HYDRAULIC INLET PORT — Hydraulic Piston — Push Rod

....
....
....
....
....
....
....
....

CLUTCH STOP

The purpose of a clutch stop, or brake, is to prevent the centre plate and primary shaft assembly from spinning or rotating too long following clutch disengagement.

On which vehicles are clutch stops used and why?

....
....
....

Operation (clutch stop)

A clutch disengagement mechanism is shown below. Add to it by sketching a simple clutch stop and describe its operation.

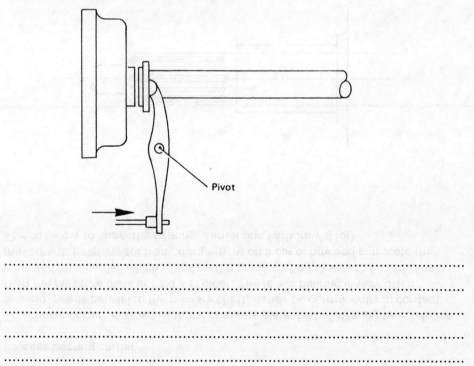

Pivot

....
....
....
....

RELEASE BEARING

The release bearing transmits the axial thrust to the release levers, plate or diaphragm or the clutch pressure plate during engagement and disengagement.

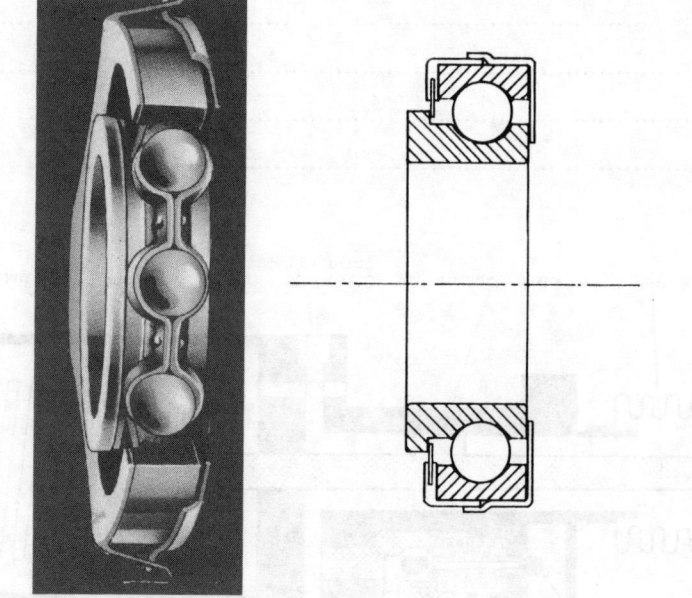

The bearing shown is a single row deep groove ball bearing. This type of bearing will withstand the axial thrust imposed during operation.
How is this type of release bearing lubricated?

....
....
....

State the purpose of the ball bearings?

....
....
....

Release bearing carrier

To achieve correct contact with the pressure plate the ball bearing type release bearing moves parallel to the primary shaft; it may be permanently in contact with the pressure plate or held just clear. The release bearing is very often mounted on a carrier which slides on the primary shaft sleeve to provide linear movement. Examine such an arrangement on a car or hgv and complete the sketch below to show the bearing, carrier and withdrawal fork.

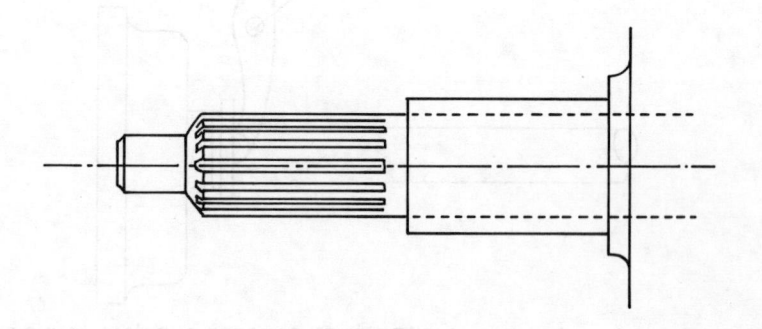

Spigot bearing

The front end of the first motion shaft is supported in a bearing located in the flywheel; name two types of bearing used for this purpose and add one to the drawing above.

....

How is a spigot shaft bush in the flywheel lubricated?

....
....

MULTI-PLATE CLUTCHES

The multi-plate clutch is a clutch assembly which utilises more than one clutch plate.

There are three basic reasons for using multi-plate clutches:

1. *Allows a reduction in clutch diameter*

   ....

2. ....

3. ....

Complete the drawing below of a multi-plate clutch.

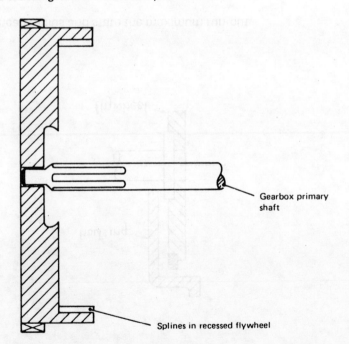

Gearbox primary shaft

Splines in recessed flywheel

State the purpose of the splines in the flywheel.

....
....

A typical example of a hgv multi-plate clutch is the twin disc type shown below. Complete the labelling on the drawing and state the purpose of the intermediate plate and drive pins.

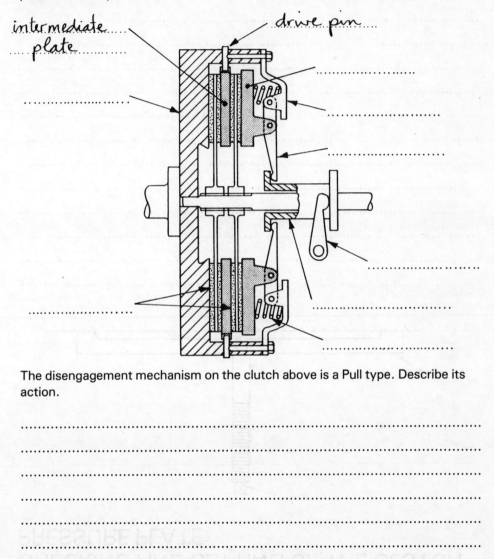

intermediate plate

drive pin

The disengagement mechanism on the clutch above is a Pull type. Describe its action.

....
....
....
....
....
....

Show on the drawing below how the flywheel should be checked for run-out.

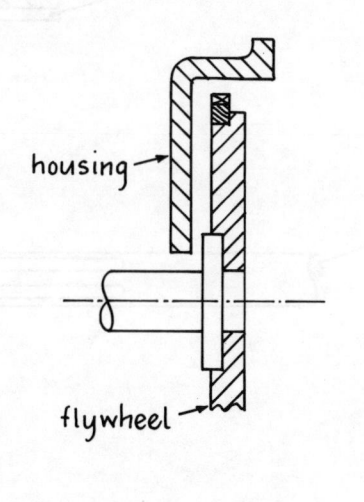

housing

flywheel

Indicate the mean radius and state the maximum run-out.

Maximum run-out ....

A hydraulically operated clutch needs to be bled on certain occasions. When are these and what equipment is needed?

....
....
....

A clutch service arbor and table are shown top right. This equipment, together with a press, is used when carrying out repairs and adjustments to a clutch pressure plate. Complete the drawing to show how the equipment would be used, and describe the adjustments and measurements to be made.

CHECKING AND SETTING UP THE CLUTCH PRESSURE PLATE

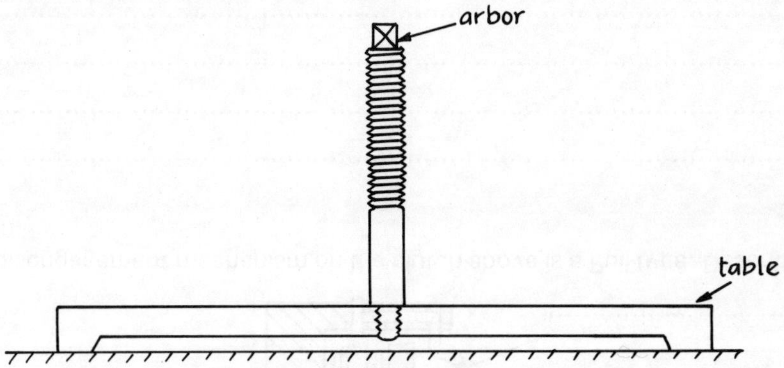

arbor

table

....
....
....
....
....
....
....
....
....
....
....
....
....
....
....
....

CLUTCH SYSTEM AND COMPONENT PROTECTION

How might the system and its components be protected during use or repair against the hazards listed?

(a) Ingress of dirt ....

....

....

(b) Moisture absorption by fluid ....

....

....

(c) Mixing of incompatible fluids ....

....

....

(d) Fluid contamination ....

....

....

(e) Friction face contamination ....

....

....

(f) Excessive facing wear ....

....

....

(g) Shock loading ....

....

....

(h) Undue cable and pivot wear ....

....

....

CLUTCH MAINTENANCE ADJUSTMENTS

Preventive routine maintenance is necessary in order to ensure clutch efficiency, prolong its life and minimise the risk of failure. List six important maintenance tasks.

....

....

....

....

....

....

....

Outline clutch tests in respect of the following:

Engagement and disengagement ....

....

....

....

Abnormal noise ....

....

....

Abnormal vibration ....

....

....

Clutch brake operation ....

....

....

Complete the table below for the clutch faults listed.

Complete the table below in respect of general rules for efficiency and precautions to be observed during clutch maintenance and repair.

FAULT	SYMPTOMS	CAUSES
Slip		
Drag		
Fierceness or Snatch		
Judder		
Squeak or Rattle		
Spin		

OPERATION	GENERAL RULES
Lifting and supporting. Preventing distortion	
Obtaining correct free play	
Correct fitting of centre plate	
Ensuring component cleanliness	
Avoiding fluid spillage. Disposal of waste.	
Use of clean fluid.	

19.3 TORQUE AND POWER TRANSMITTED BY FRICTION CLUTCHES

The factors affecting the torque transmitted by a friction clutch are:

the coefficient of friction between the contact surfaces μ

the total force exerted by the pressure plate springs W

the mean radius of the clutch linings r

the number of pairs of contact surfaces n

These values are multiplied together to obtain the torque transmitted by a friction clutch, that is

torque transmitted $= \mu W r n$

What is the mean radius of a clutch plate whose friction rings are 0.25 m outside diameter (r_1) and 0.15 m inside diameter (r_2)?

Mean radius (r) $=$ _____ $=$ .... m

How many pairs of contact surfaces has a single-plate clutch? ....

Having found the torque, if the speed of rotation of the clutch (N) is known it is a simple matter to calculate the power transmitted by the clutch, using the formula:

Power $= 2\pi N T$

INVESTIGATION

Conduct a coefficient of friction experiment using various sections of clutch lining, and state the coefficient of friction for:

(a) a new clutch lining ....

(b) a polished clutch lining ....

(c) a glazed clutch lining ....

INVESTIGATION

To show the effect of the number of pairs of contact surfaces:

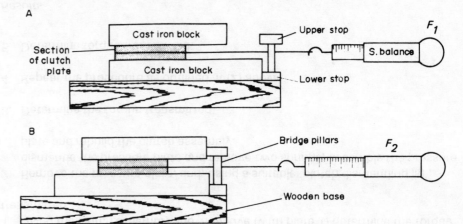

1. Using apparatus similar to that shown above, determine the force required to slide the friction material on lower CI block while carrying the upper CI block (A above).

2. Note force required to slide friction material between CI blocks (B above).

   Spring balance reading before contacting stop (F_1) $=$
   Spring balance reading after contacting stop (F_2) $=$

What effect have the number of contact surfaces on the force of friction?

....

....

How can this effect be used to advantage in a clutch assembly?

....

....

INVESTIGATION

Effect of mean radius on torque transmitted by a clutch:

1. Using a small flywheel and clutch assembly mounted as shown opposite (or any similar arrangement), determine the torque necessary to turn the clutch plate, that is, slip the clutch.

2. Remove the clutch plate and refit an identical plate which has had the outer part of the lining reduced; see diagram below.

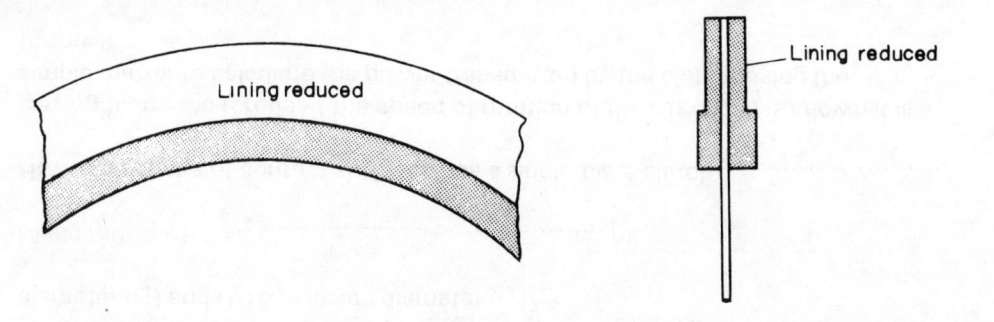

3. Again use the torque wrench to determine the torque transmitted by plate 2.

4. Mean radius of plate 1 = ...

Torque transmitted by plate 1 = ...

Mean radius of plate 2 = ...

Torque transmitted by plate 2 = ...

5. State the effect of fitting a friction lining which has the same overall diameter but larger area.

....
....
....

INVESTIGATION

Effect of spring strength on torque transmitted by a clutch:

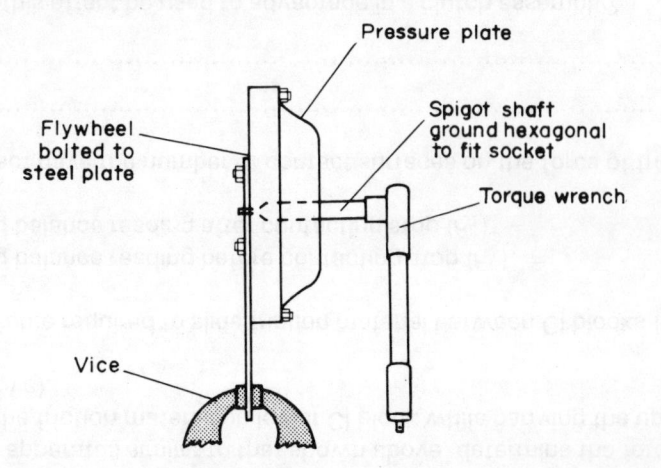

1. Using the clutch assembly shown above (with plate 1) determine the torque transmitted.

2. Remove the pressure plate, and, using a suitable 'clutch dismantling jig', dismantle the pressure plate and remove two springs; assemble the pressure plate and rebuild the clutch assembly.

3. Determine the torque transmitted.

4. Repeat the procedure as outlined at (2) above.

5. Determine torque transmitted.

Results

Torque transmitted using all six springs ....

Torque transmitted using four springs ....

Torque transmitted using two springs ....

16

Problems

1. A single-plate clutch has a coefficient of friction of 0.3 and the friction linings have a mean radius of 75 mm. If the total force exerted by the pressure plate springs is 2200 N, calculate the torque transmitted by the clutch.

   Torque transmitted $= \mu Wrn$

   torque transmitted $= \dfrac{0.3 \times 2200 \times 75 \times 2}{1000}$

   $\therefore$ torque transmitted $= 99$ Nm *Ans.*

2. A single-plate clutch has facings of 225 mm outside diameter and 135 mm inside diameter and a coefficient of friction of 0.24. The nine pressure plate springs each exert a force of 320 N. Calculate:

   (a) the maximum torque that could be transmitted by the clutch.

   (b) the power transmitted by the clutch at a speed of 50 rev/s.

   ....

   ....

   ....

   ....

   ....

   ....

   ....

   ....

   ....

3. Determine the coefficient of friction between the contact surfaces of a single-plate clutch which will transmit a torque of 125 Nm, if the total spring force acting on a friction plate of 100 mm mean radius is 2500 N.

   ....

   ....

   ....

   ....

   ....

   ....

4. A single-plate clutch has a mean radius of 80 mm and a coefficient of friction of 0.3. What force must be exerted by each of six springs to enable the clutch to transmit a torque of 105 Nm?

   ....

   ....

   ....

   ....

   ....

5. A multi-plate clutch has three friction plates of 60 mm mean radius running in oil. The coefficient of friction is 0.15 and the total spring force is 1000 N. Calculate the torque and power transmitted by the clutch at 40 rev/s.

   ....

   ....

   ....

   ....

Calculations (clutch linkage)

Two types of lever used in clutch linkages are shown below. In each type of lever the load and effort will each produce a 'clockwise' and 'anti-clockwise' turning moment about the fulcrum or pivot. The turning moment is the **product** of the *force* and the *perpendicular distance from the fulcrum*.

When a state of balance (equilibrium) is maintained:

Clockwise moments = anti-clockwise moments

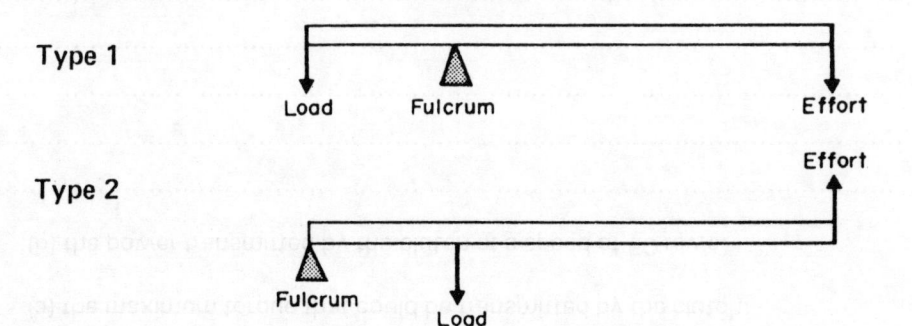

Type 1

Load   Fulcrum   Effort

Type 2

Fulcrum   Load

1. Calculate the pull on the cable for the pedal lever shown when the force exerted at the pedal is 90 N.

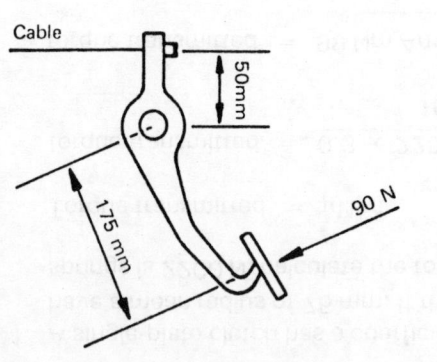

Cable

50mm

175 mm

90 N

2. Calculate the force exerted by the slave cylinder push rod below when the force at the release bearing is

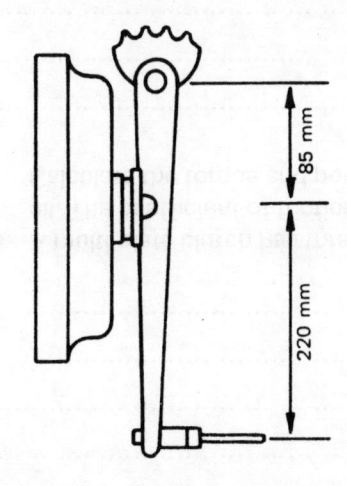

85 mm

220 mm

In question (1), owing to the ratio of the lever, the movement at input (pedal) is 3½ times greater than the movement at output (cable)
This is known as MOVEMENT RATIO.

MOVEMENT RATIO = ...

FORCE RATIO = ... or ....

% EFFICIENCY = ... × 100

3. Calculate the force required at right angles to the clutch pedal lever to operate the clutch with the mechanism shown below.

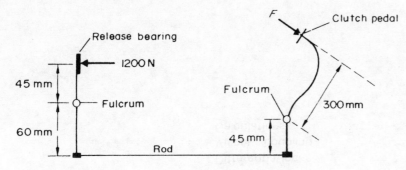

5.

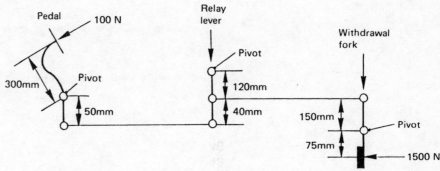

Calculate the MR, FR and EFFICIENCY for the hgv clutch-operating system shown above.

4. In a cable-operated clutch a pedal effort of 90 N is applied through a distance of 150 mm. If the force applied at the release bearing is 540 N, calculate MR, FR and EFFICIENCY for the system if the movement of the release bearing is 20 mm.

6. In a hgv the force required at the release bearing to disengage the clutch is 2400 N. Calculate the pedal effort if the clutch operation system has a movement ratio of 18:1 and an efficiency of 90%.

Chapter 2

Manual Gearboxes

ELEMENTS 20/37/38   **UNITS 16/35/36**

MANUAL GEARBOXES

Location

In a vehicle transmission the gearbox is located between the clutch and final drive gears. Its actual location on the vehicle, however, depends on the layout of the main components. Make simple line diagrams to show the layout for types indicated on this page.

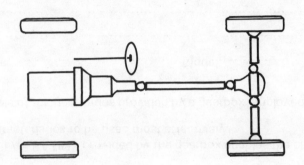

FRONT ENGINE REAR WHEEL DRIVE

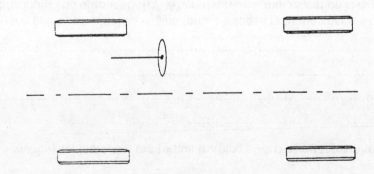

FRONT ENGINE FRONT WHEEL DRIVE

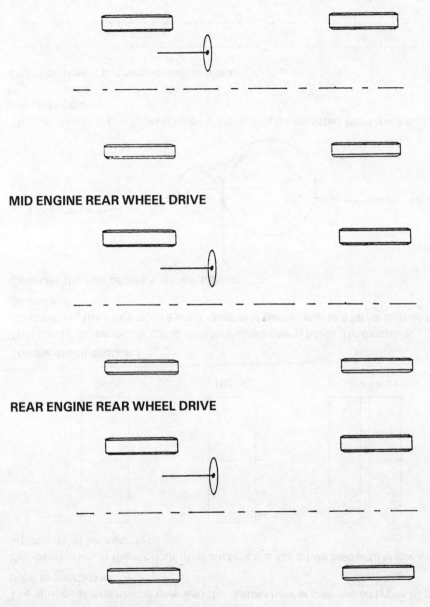

MID ENGINE REAR WHEEL DRIVE

REAR ENGINE REAR WHEEL DRIVE

REAR ENGINE REAR WHEEL DRIVE (psv)

FUNCTION

Basically the gearbox serves a number of purposes:

1. To multiply (or increase) the torque (turning effort) being transmitted

....

....

....

....

When the gearbox multiplies engine torque a speed reduction occurs between gearbox input and output shafts. When the torque multiplication is high the vehicle speed is .... The gear in which maximum torque multiplication is obtained is ....

The torque output from the gearbox is therefore varied according to the SPEED AND LOAD REQUIREMENT of the vehicle.

The range of GEAR RATIOS provided by the gearbox enables the engines torque and speed characteristics to be used most effectively.

Give an example of the gear ratios provided by a gearbox employed in a modern car or hgv.

Vehicle's make .... Model ....

....

....

....

Modern hgv gearboxes provide a range of gear ratios far in excess of those available to the car driver. Why is this?

....

....

....

Types of gearing

The spur gear, the helical gear and the double helical gear are all types of gears used in gearboxes.

By observation in the workshop, complete the sketches below to show the tooth arrangement for each type.

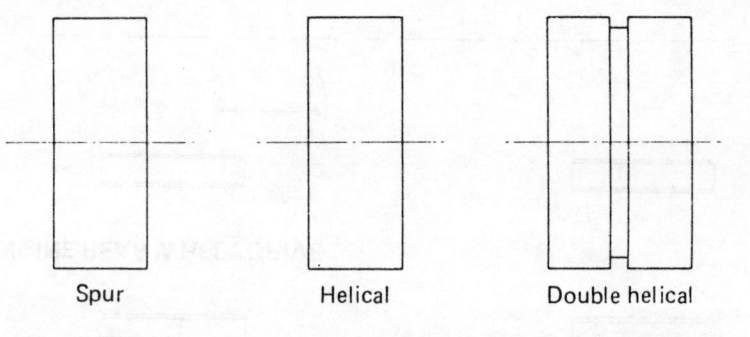

Spur   Helical   Double helical

Torque multiplication

This can be achieved by using different-sized gearwheels, for example, transmitting the drive from a small-diameter gearwheel to a large-diameter gearwheel.

Consider the pair of gears shown below:

F

Input torque ———   ——— Output torque

r R

The input torque produces a force (F) which is transmitted from the small gear to the large gear.

Describe how input torque is multiplied.

....

....

....

CONSTANT-MESH GEARBOX

In the constant-mesh gearbox, as the name implies, the gearwheels are permanently in mesh. Constant meshing of gears is achieved by allowing the mainshaft gearwheels to rotate on bushes. Dog clutches, which are splined to the mainshaft, provide a positive connection when required, to allow the drive to be transmitted to the output shaft.

Principle of operation:

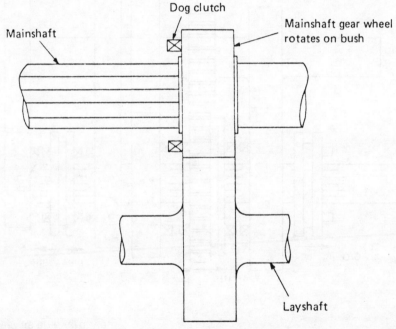

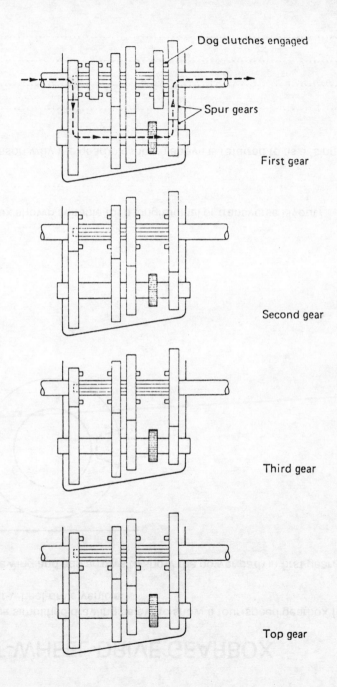

Dog clutches engaged

Spur gears

First gear

Second gear

Third gear

Top gear

Power flow

(a) Complete the drawing above to show how a dog clutch member is used to connect the mainshaft gearwheel to the mainshaft and add arrows to indicate the power flow from the layshaft to the mainshaft.

(b) Add arrows to the drawings opposite to trace the power flow through the constant mesh gearbox in the gear positions shown. Sketch the dog clutches in their correct positions.

FIVE-SPEED GEARBOX

In most four-speed gearboxes the top ratio is 1:1, that is, the primary shaft and mainshaft are connected to give a direct drive through the gearbox. Many modern vehicles, however, are fitted with a five-speed gearbox in which top gear is an 'overdrive' or 'step-up' gear.

Complete the drawing below to show a five-speed constant-mesh gearbox with top gear as an overdrive.

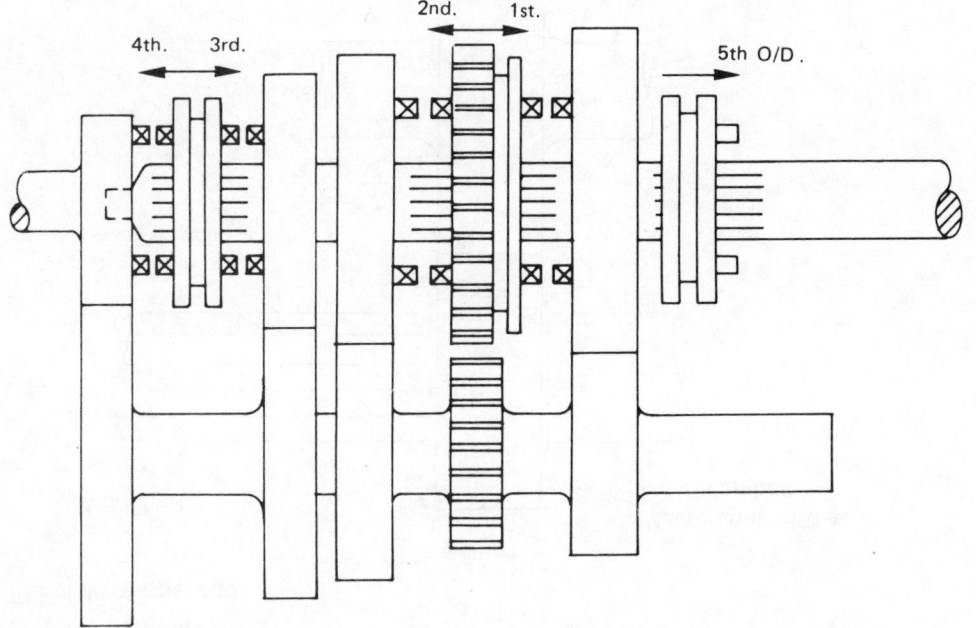

Show, on the drawing above, how reverse gear is obtained.

....

....

....

FRONT-WHEEL-DRIVE GEARBOX

Complete the simplified drawing below to show a four-speed gearbox for a front engine, front-wheel-drive vehicle.

Label the drawing and add arrows to show the power path in first gear.

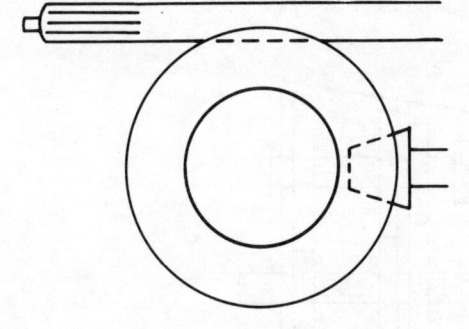

Is the gearbox shown suitable for a longitudinal or transverse layout?

....

State the reason why the gearbox shown above is referred to as a 'single-stage' gearbox.

....

....

....

20.4 SYNCHROMESH DEVICES

When a gear-change is made, the speeds of the meshing dog clutch teeth must be equalised in order to avoid clashing which would result in noise and wear. As the name implies, the synchromesh device synchronises the speeds of the dog teeth during gear-changing.

Complete the drawing at B opposite to show third gear engaged and describe (in the space below) the operation of the unit.

....
....
....
....
....
....
....
....
....
....
....

State the factors which affect the frictional force on the cones.

....
....
....
....
....
....

Constant-load synchromesh device

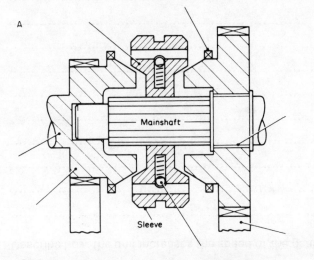

A

Mainshaft

Sleeve

Complete the labelling on this drawing.

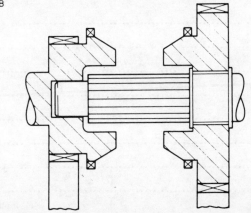

B

State the reason why this device is known as the 'constant-load' type.

....
....
....

25

Baulk-ring synchromesh

With the constant-load type of synchromesh, if a quick gear-change is made the dog clutches can come into contact before their speeds are properly synchronised. This causes noise and wear, that is, it is possible to 'beat' the synchromesh.

The baulk-ring synchromesh is a development of the constant-load type. The characteristics of the baulk-ring type are that:

(a) the synchronisation of the dog clutches is quicker, thus allowing a quicker gear-change;

(b) the dog clutches cannot be engaged until their speeds are equal.

The sketch at A represents a section of a baulk-ring synchormesh unit and the mating dog clutch.

Examine a baulk-ring synchromesh unit and complete the sketch B to show how the baulk ring prevents movement of the outer sleeve until synchronisation has taken place.

A Gear   Dog teeth   Outer sleeve   B

Slot

Baulking ring   Shifting plate

Briefly describe the operation of the baulk-ring type of synchromesh unit, making particular reference to the function of the baulk-ring and the shifting plates.

....

....

....

....

....

....

....

....

....

....

....

Describe how the unit increases the speed of the gear-change.

....

....

....

....

....

....

....

....

....

....

....

SELECTOR MECHANISMS

When a gear is selected the movement of the gear lever is transferred to the synchro-hub or gear-wheel through a selector fork. The fork may be fixed to a sliding shaft, or the fork may slide on a fixed shaft. Provision is also made for locking the selector fork in the required gear position and also for preventing movement of more than one selector at a time.

Examine different types of selector mechanism and complete the simplified drawing below to include the selector fork and gear lever.

Simple selector mechanism (side view)

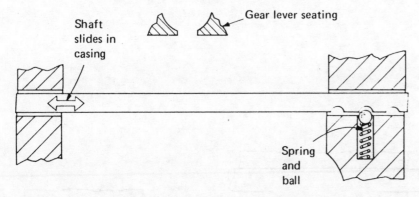

Shaft slides in casing

Gear lever seating

Spring and ball

An exploded view of a selector mechanism is shown below; label the drawing.

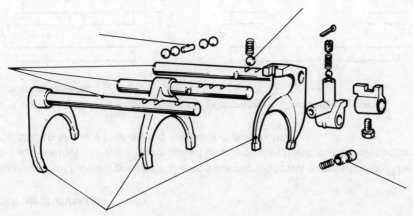

INVESTIGATION

Examine a 'sectioned' gearbox which has a DIRECT ACTING gear shift.

State:

1. The number of gears ....

2. The number of selector rails and forks ....

Show by sketching below:

(a) how accidental selection of reverse gear is prevented;

(b) how the REVERSING LIGHT switch is operated.

Interlock mechanisms

The interlock mechanism prevents the engagement of two gears at once. The drawing at A below shows a three-speed interlock mechanism. Complete the drawing at B to show a four-speed interlock mechanism.

Examine various gearboxes and show by sketching at C one other form of interlock mechanism.

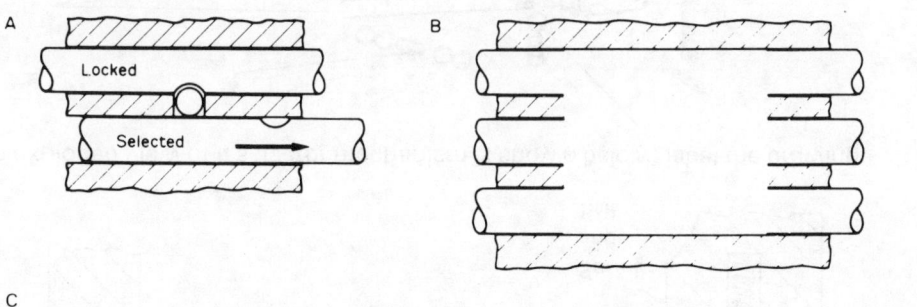

C

Remote control gearshift systems

Give reasons for using remote control systems.

....

....

....

....

....

The mechanical remote control shown below allows the use of a short gear lever.

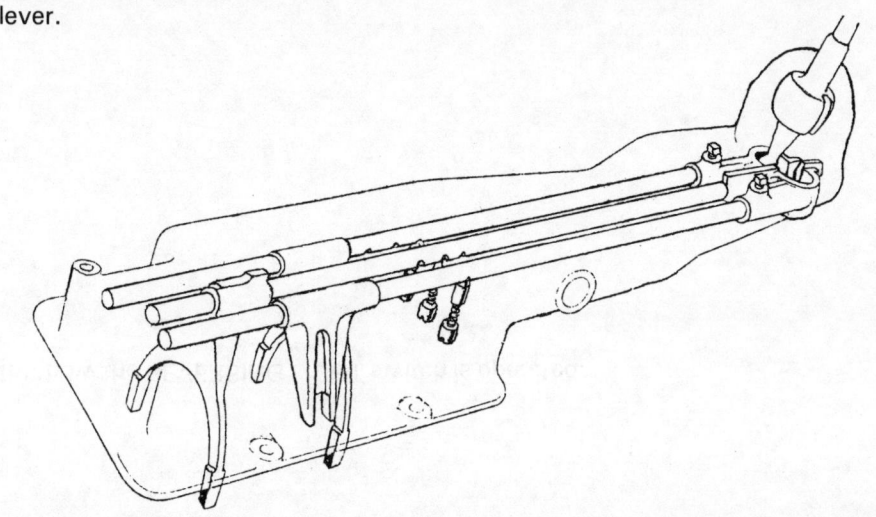

Study an alternative remote control system, such as electrical, pneumatic, etc., on a vehicle or by use of manuals, trade journals etc. and with the aid of a simple diagram describe the system.

Gearbox bearings

By means of arrows on the drawing opposite indicate the location of all the
gearbox bearings in this hgv gearbox.
Complete the labelling by naming the bearings and, below the drawing, give the
main reasons for using the particular bearings in each location.

Gearbox mountings

The flexible mountings through which the gearbox is attached to the vehicle
frame reduce noise and prevent vibration being transmitted to the vehicle
structure. Examine a vehicle and make sketches below to illustrate the location
and type of gearbox mountings employed.

Make .... Model ....

....
....
....
....
....
....
....
....
....
....
....
....
....

Oil sealing

Name the two oil seals shown below and give examples of their uses in manual and automatic gearboxes.

....

....

....

....

....

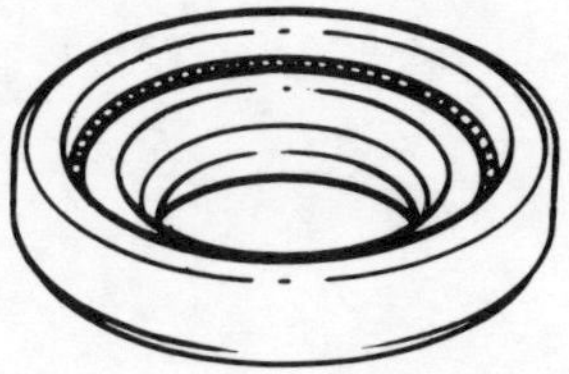

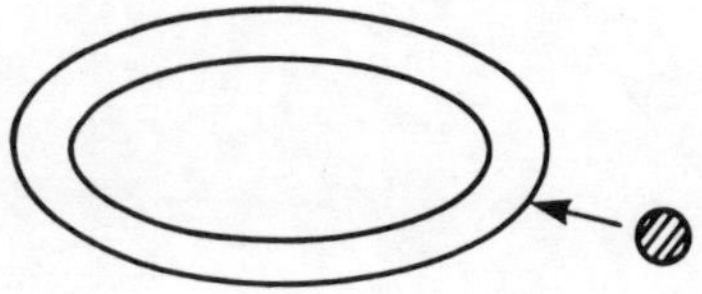

Add an oil return scroll and slinger washer to the gearbox primary shaft shown below and describe the action of these oil-retaining devices.

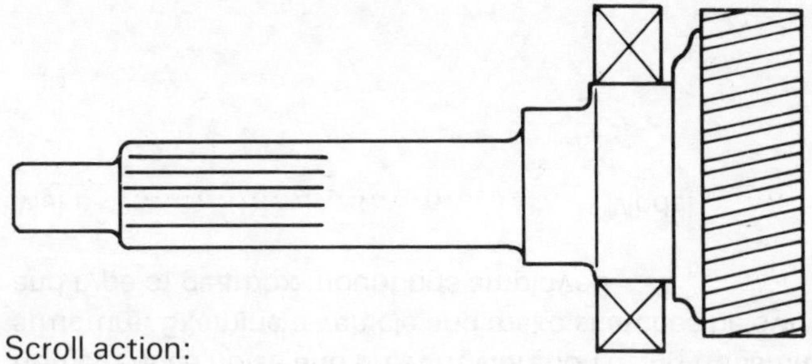

Scroll action:

....

....

Slinger washer:

....

....

....

GEARBOX AND LINKAGE ROUTINE MAINTENANCE AND LUBRICATION

Regular maintenance and proper lubrication are essential to extend gearbox life, improve efficiency and minimise failure risk of components.

List SIX maintenance items applying to gearbox and linkage.

....

....

....

....

....

....

Transmission lubricants

The main functions of a gearbox lubricating oil are:

....

....

....

....

State the reason why different lubricants are used in manual and automatic gearboxes.

....

....

....

....

....

AUXILIARY GEARBOXES (hgvs)

It is now common practice for heavy goods vehicles to employ ten or more gear ratios. One method of increasing the number of gear ratios is to operate an auxiliary gearbox in conjunction with a five- or six-speed gearbox.

Two types of auxiliary gear arrangements in use are:

SPLITTER GEARBOXES and ....

State the reason why one gearbox is referred to as the 'splitter' type.

....

....

....

....

....

....

....

....

A splitter-gear arrangement is shown opposite in tandem with a five-speed gearbox. Describe, below the drawing, the operation of the gearbox, including gear-selection procedure.

Describe how the splitter gear is usually controlled from the driving position.

....

....

....

....

The splitter gearbox

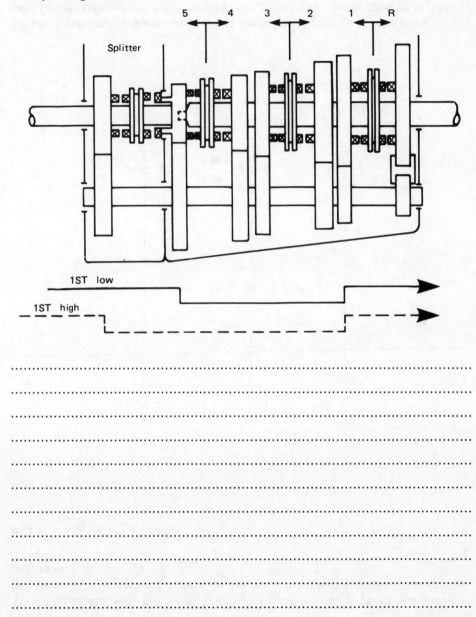

....

....

....

....

....

....

....

....

....

Range change gearbox

The range change is an auxiliary gearbox, usually attached to the rear of the main gearbox, which provides a low range of gear ratios and a high range. With this gearbox the driver would change as normal through the gears from first to fifth, with the auxiliary gear in low. He would then engage first gear again, but this time in high ratio, that is, 'straight through' the auxiliary gearbox, and gear-change as normal up to fifth gear, thus obtaining ten speeds.

Add a range change section to the rear of the five-speed gearbox shown below and indicate the power flow for high and low ratio.

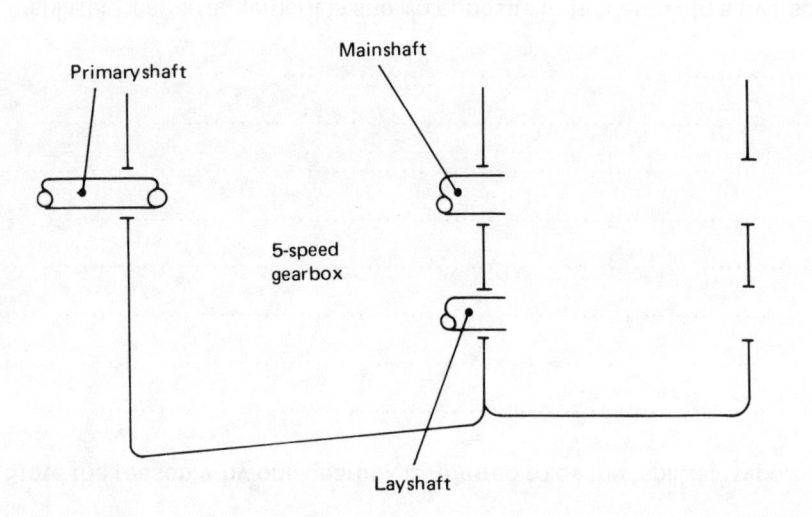

Primary shaft

Mainshaft

5-speed gearbox

Layshaft

The shift pattern, with typical gear ratios, for a range change gearbox is shown below. From the information given on the drawing state the auxiliary reduction gear ratio.

Auxiliary reduction gear ratio: ....

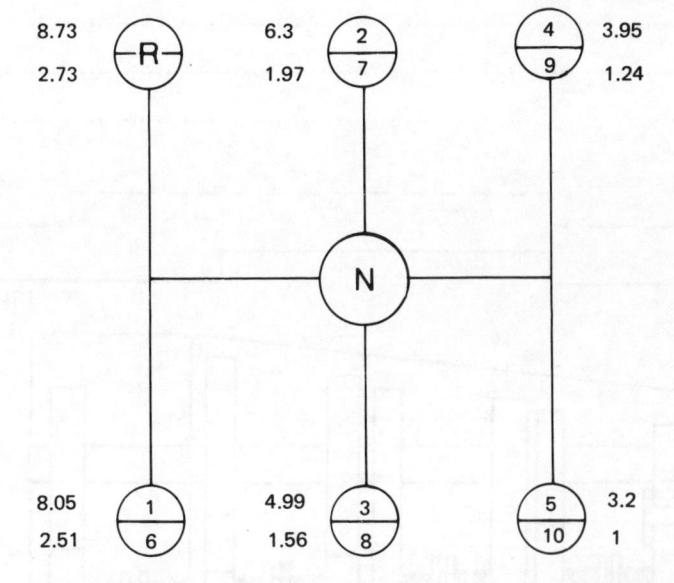

Although ten gear ratios are available to the driver, he would not necessarily select every ratio up and down through the gears. It is usual procedure to 'skip' gears and select the gear appropriate to the speed and load of the vehicle.

TWIN LAYSHAFT GEARBOX

An extremely popular hgv gearbox is the twin layshaft or twin counter-shaft type. In this design the drive from the primary gear is transmitted to the mainshaft gears via two layshafts which are positioned on either side of the mainshaft.

The simplified drawing below shows the mainshaft assembly, which is said to be 'floating', of a ten-speed twin layshaft transmission. Complete the drawing by adding the layshafts, including range change section, and the labelling.

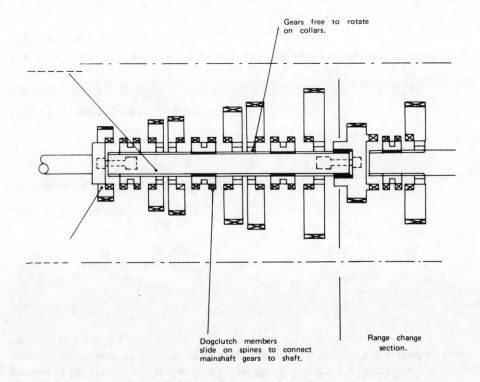

Gears free to rotate on collars.

Dogclutch members slide on spines to connect mainshaft gears to shaft.

Range change section.

Torque balance

The upward-tooth load on one side of the mainshaft gears is balanced by the downward-tooth load on the opposite side; the mainshaft gears do therefore 'float' between the layshaft gears, that is, the mainshaft gears are located radially by the layshaft gears.

Describe how the tooth-load balance or torque balance is achieved.

....

....

....

....

....

....

State the advantages of the twin layshaft type of gearbox compared with the single layshaft type.

....

....

....

....

....

....

....

....

....

....

....

....

33

Speedometer/tachograph drive arrangement

The speedometer or tachograph is usually driven by a gear arrangement (shown below) at the rear end of the gearbox mainshaft. The driving gear shown is a push fit on the rear of the mainshaft and is clamped between the mainshaft rear bearing and the U/J flange.

(a)   (b)

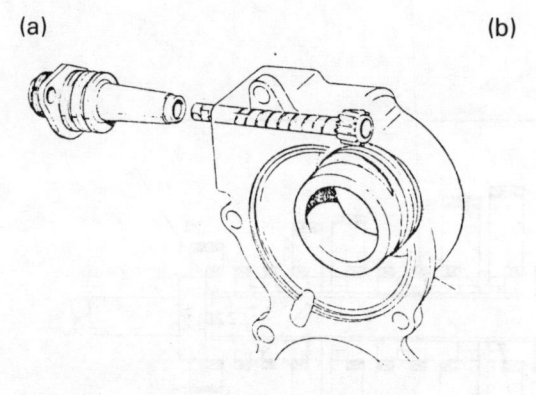

Name the type of gears shown above.

....

Make a sketch at (b) to show an adaptor gearbox in the driveline to a tachograph and state the purpose of this unit.

....

....

....

State the statutory requirements relating to the calibration and sealing of tachographs.

....

....

....

....

Indicate on the gearbox shown below the location of oil seals, gaskets, level plug and vent.

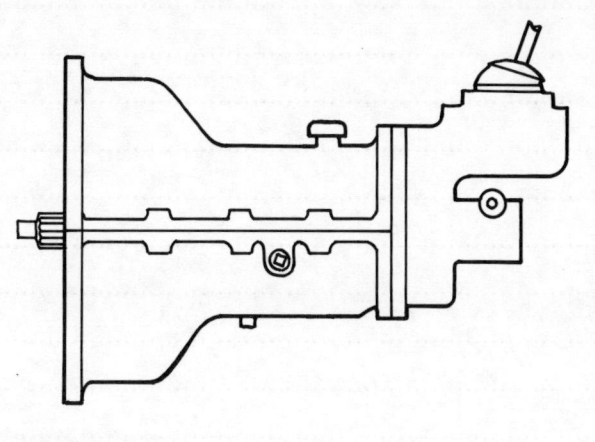

State the purpose of:

Gaskets

....

....

Level plug

....

....

Vent

....

....

State the quantity and type of oil used to lubricate a typical car and hgv gearbox.

Car make .... Model ....

Oil quantity .... Type ....

hgv make .... Model ....

Oil quantity .... Type ....

Gearbox faults

Complete the table below to show common gearbox faults associated with the symptoms listed.

Symptoms	Faults
Noisy operation	
Difficulty in obtaining a certain gear	
Jumping out of gear	
'Sloppy' gear lever action	
Oil leakage from rear of gearbox	
Regular ticking or knocking noise	
Gearbox 'locked up solid', that is shafts will not rotate	
Inoperative speedo/tacho	

TESTING AND TEST EQUIPMENT

Problems associated with the gearbox become evident during testing and checking procedures which include:

  Road testing
  Operating the vehicle on a rolling road dynamometer
  Visual external examination of the gearbox, linkage
  and mountings
  Examination of the oil.

Carry out diagnostic tests on a gearbox either by operation on road or dynamometer or on a partially dismantled gearbox. List the equipment used and faults found.

Vehicle make .... Model ....

EQUIPMENT

....

....

....

FAULTS

....

....

....

....

Describe the operation and maintenance of ONE item of test equipment.

....

....

....

....

....

INVESTIGATION
GEARBOX AND OPERATING LINKAGE
REMOVAL AND REFITTING

Gearbox

As a result of workshop experience, or by reference to a workshop manual, describe the procedure for removing a typical gearbox (for linkage, see opposite).

Vehicle make .... Model ....

Gearbox type ....

....

....

....

....

....

....

....

....

....

....

....

....

....

....

....

....

Linkage

Additional to the gearbox removal, describe how to remove, replace and adjust the gear/change linkage.

....

....

....

....

....

....

....

....

....

....

While carrying out gearbox work, how might the box and linkage be protected against the following hazards?

Oil leakage ....

....

....

Ingress of dirt and moisture ....

....

....

Linkage seizure ....

....

....

Pressure build up ....

....

....

POWER TAKE OFF (PTO) SYSTEMS

A PTO enables a wide range of installed equipment, towed and standing machinery to be driven by the vehicles engine, for example, generators, pumps, winches etc. Dual purpose vehicles used by farmers and construction workers and specialist vehicles (Volvos shown) use PTOs.

Name four power take off points for the PTO.

1. *Gearbox mainshaft*   2. ....

3. ....   4. ....

Examine an alternative to the centre PTO arrangement shown opposite and make a simple line drawing below to illustrate it.

In what type of transmission is the centre PTO used and how is it operated by the driver?

Selector mechanism

Power take-off control lever

Sleeve splined onto output shaft

Seal

Power output flange

Gearbox mainshaft

Bush

Power take-off unit flanged onto rear of main gearbox

....

....

....

....

The PTO above is selected mechanically. Describe with the aid of a simple line diagram an alternative system of selection, for example, electrical, pneumatic etc.

....

....

....

....

OVERDRIVE

An overdrive exists when the input shaft or gear rotates more slowly than the output shaft or gear.

If a simple pair of gears were used to transmit the drive from input to output shaft, overdrive would occur if the input gear was the larger gear.

State the advantages of this type of overdrive.

....

....

....

....

A simplified drawing of a separate overdrive unit, normally mounted on the rear of the gearbox, is shown opposite; complete the labelling on the drawing.

The drawing shows the unit in 'direct drive'; describe its operation in this position.

....

....

....

....

....

Describe how the drive is transmitted in reverse and during overrun.

....

....

....

....

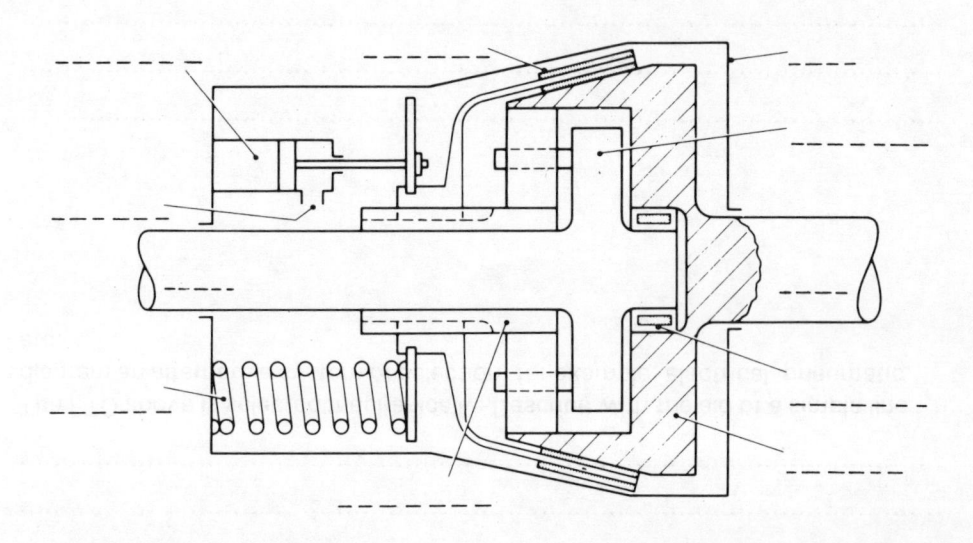

INVESTIGATION

Examine a dismantled overdrive unit and complete the simple sketches below to illustrate the action of the unidirectional clutch.

Drive   Freewheel

....

....

....

....

Complete the drawing to show the overdrive engaged and indicate on the drawing the power path through the unit. Describe below the drawing the operation of the unit when in overdrive.

The overdrive is operated via an electric solenoid controlled by a switch. Complete the drawing below to show the complete electrical circuit for overdrive control.

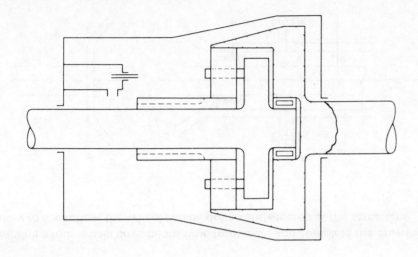

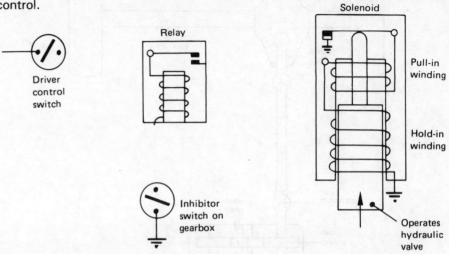

Operation

....
....
....
....
....
....
....

Other than when in direct drive, state one other operating condition during which the unidirectional clutch is maintaining drive.

....
....

Describe below, with the aid of a simple diagram, the action of the PUMP and ACCUMULATOR.

....
....
....
....
....
....

FOUR-WHEEL DRIVE (4WD or 4 × 4)

Four-wheel drive vehicles are either:

(a) Heavy truck type vehicles operating for part of the time off the normal road on uneven, soft or slippery surfaces; or

(b) ....

(c) ....

With the simple system shown, four-wheel drive is engaged for use on uneven, soft or slippery surfaces. It should be disengaged for normal road use. Why is disengagement necessary?

....

....

....

....

Examine a four-wheel drive beam-type front axle, and complete the drawing below to show how the wheel can be driven and steered at the same time.

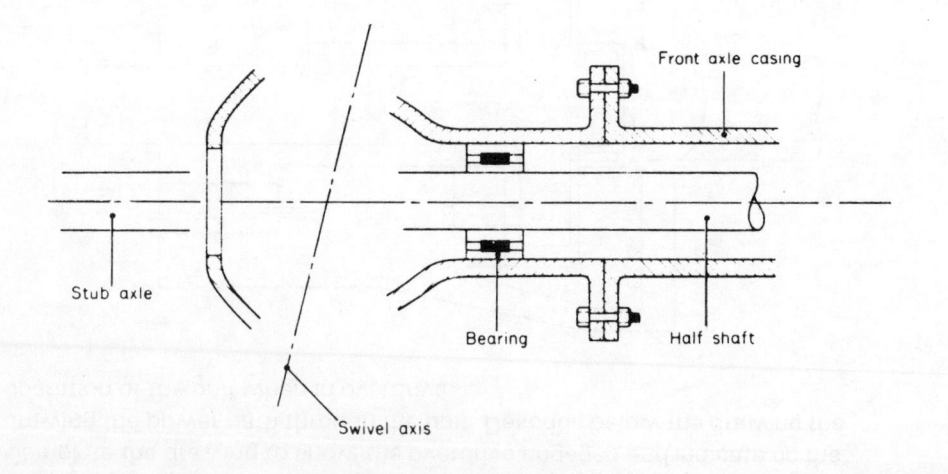

Stub axle

Swivel axis

Front axle casing

Bearing

Half shaft

Layout

Complete the drawing below to show the front and rear final drive gears and show how the drive to the front and rear axle can be disconnected.

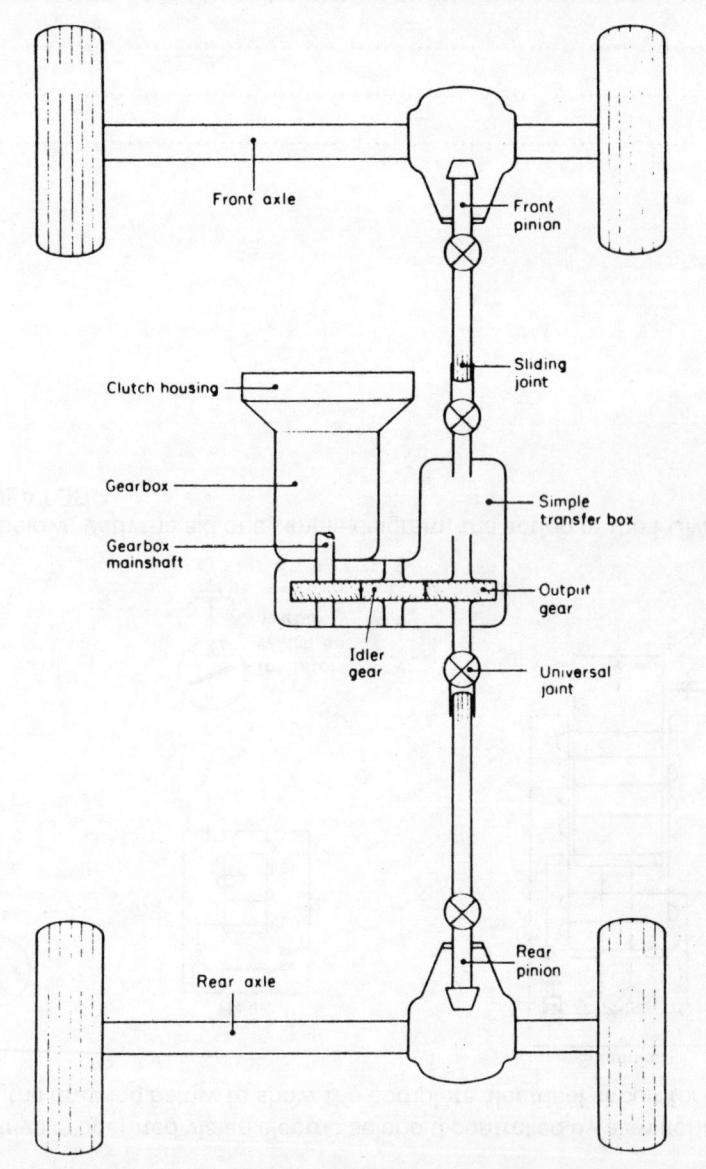

Front axle

Front pinion

Sliding joint

Clutch housing

Gearbox

Gearbox mainshaft

Simple transfer box

Output gear

Idler gear

Universal joint

Rear axle

Rear pinion

TWO-SPEED TRANSFER BOX

Complete the drawing and labelling of the two-speed transfer box below and describe its operation in low and high ratio.

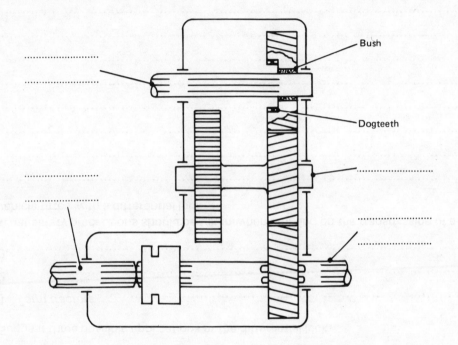

....

....

....

....

Bush

Dogteeth

....

....

Operation (two-speed transfer gearbox)

....

....

....

....

....

How does a two-speed transfer box affect the transmission gearing as a whole?

....

....

....

Complete the drawing below to show how the front and rear propeller shafts can be driven through a differential.

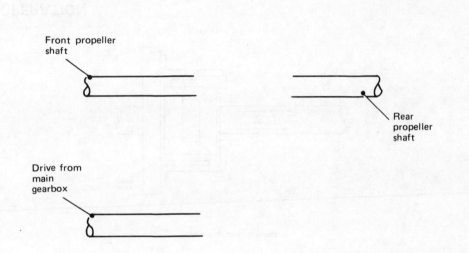

Front propeller shaft

Rear propeller shaft

Drive from main gearbox

State the reason for using a third differential:

....

....

....

To obtain the benefit of four-wheel drive when operating off the road it is usual practice to use a differential lock on the third differential.

37.1 DIFFERENTIAL LOCKS

If one driving wheel of a vehicle encounters a soft or slippery surface, the wheel will spin and the torque required to drive it will be negligible. Because of the action of the differential the other non-slipping wheel will receive the same negligible torque. The vehicle will therefore be immobilised.

The differential lock, as the name implies, locks the differential and allows maximum tractive effort allowed by the road surface to be utilised at each wheel.

Suggest three possible applications of the differential lock.

1. *Farm tractors* ....

2. ....

3. ....

What safety precautions should be taken when working on the transmission of a vehicle fitted with a differential lock?

....
....
....
....
....
....
....
....
....
....
....

Complete the sketch at 'B' below to show the differential lock engaged, and explain how the action locks the differential.

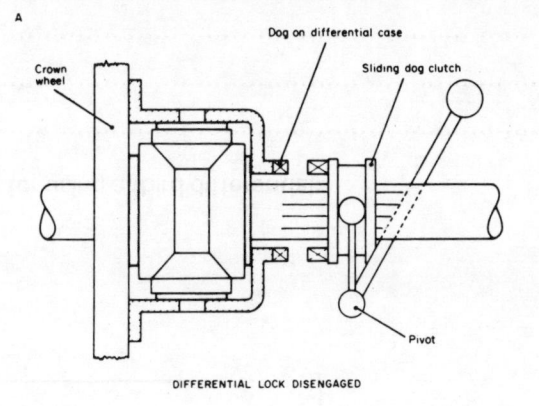

DIFFERENTIAL LOCK DISENGAGED

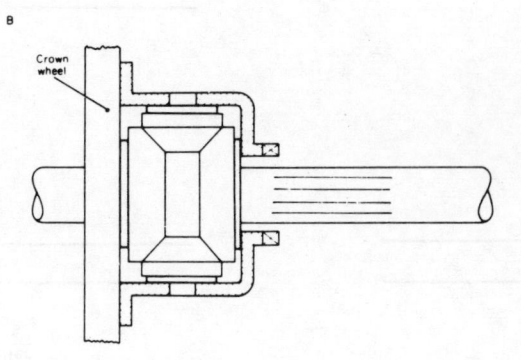

DIFFERENTIAL LOCK ENGAGED

OPERATION

....
....
....
....
....
....

VISCOUS COUPLING (VC)

Four-wheel drive operation under all conditions can be achieved by the use of LIMITED SLIP DEVICES. The system shown opposite employs viscous couplings in the centre (third) differential and in the rear differential. These couplings control wheelspin and greatly improve traction and roadholding in all drive conditions without the need for the engagement of manual differential locks by the driver.

On a car application, such as the one shown opposite, it is usual to divide the driving torque UNEQUALLY between front and rear wheels.

Give a typical percentage TORQUE SPLIT and state the reasons for this.

TORQUE SPLIT: FRONT .... REAR ....

....
....
....
....
....
....

Where in the transmission is the torque split achieved?

....

....

How is the drive transmitted to the front axle?

....

Complete the labelling on the layout shown below.

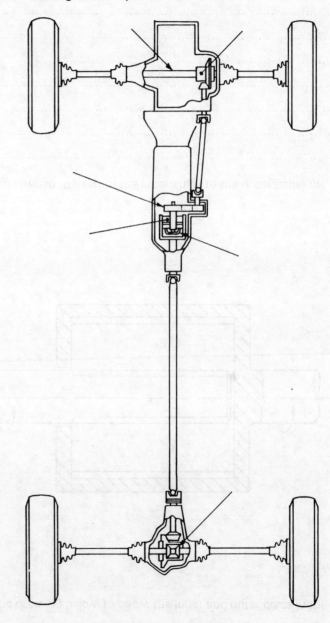

VISCOUS COUPLING – OPERATION

The structure of a viscous coupling is similar to that of a multi-plate clutch. The coupling consists of a number of INNER and OUTER discs.

OUTER DISC   INNER DISC

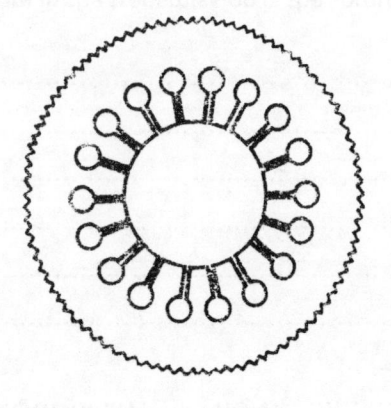 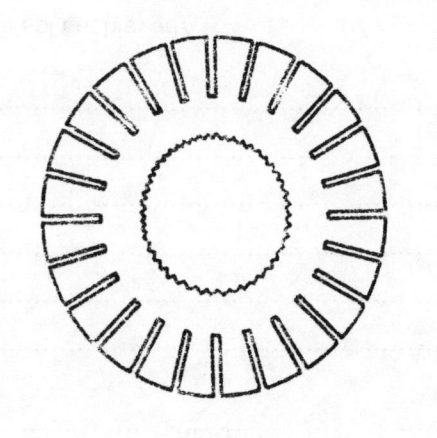

The inner discs are splined on to an inner carrier shaft and the outer discs are splined on to the inside of an outer housing. A small clearance or gap between the discs is maintained by interposed spacer rings. The gap between the discs is filled with a high viscosity silicone fluid.

Torque transmission through a VC is based on the transmission of shearing forces in the fluid.

Viscous couplings, as already stated, will control differential spin. In what other capacity are they employed in vehicle transmissions?

....
....
....
....

Complete the drawing below to show the inner and outer discs in the simplified layout.

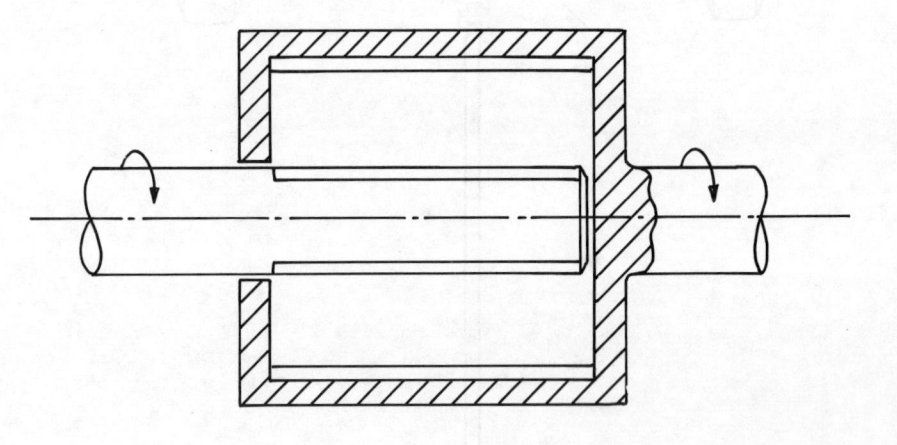

How is torque transmitted from the inner shaft to the outer housing?

....
....
....
....
....
....
....

VISCOUS COUPLING AND DIFFERENTIAL

The viscous coupling, or viscous control, when used in conjunction with a differential will limit or control spin on shafts being driver via the differential, for example, front/rear propeller shafts or drive shafts.

The simplified drawing below shows a VC incorporated into a final drive/differential unit.

Label the drawing and explain how the VC limits drive shaft spin.

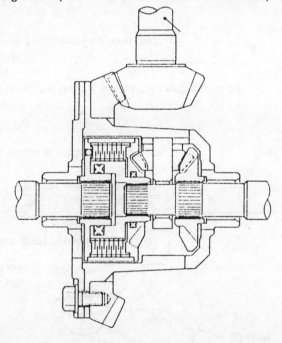

....

....

....

....

CENTRE DIFFERENTIAL

The centre differential shown below is an epicyclic gear set. Complete the drawing by adding a VC which would control spin between front and rear propeller shafts.

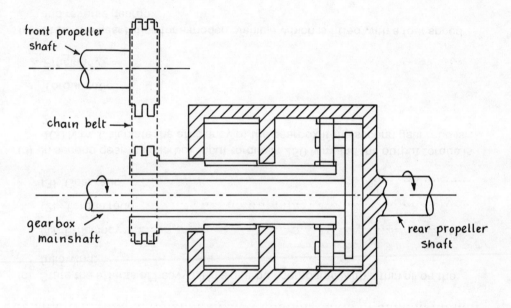

front propeller shaft

chain belt

gearbox mainshaft

rear propeller shaft

How does this differential provide the 2:1 torque split?

....

....

....

....

GEAR RATIO, TORQUE RATIO, EFFICIENCY

Gear ratio

When two gearwheels are meshed to form a 'simple' gear train, the *gear ratio* can be expressed as a ratio of gear teeth or a ratio of gearwheel speeds.

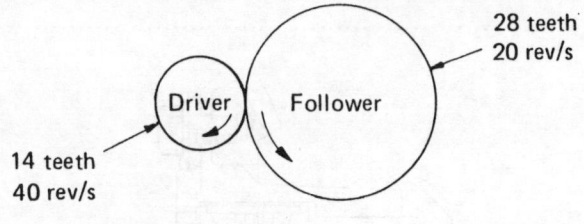

28 teeth
20 rev/s

Driver   Follower

14 teeth
40 rev/s

For the pair of gears represented above:

$$\text{Gear ratio} \;=\; \frac{\text{No. of teeth on follower}}{\text{No. of teeth on driver}} \;=$$

or using gear wheel speeds,

Gear ratio =

When speeds are used the ratio is sometimes expressed as

....

(Note: See associated problems on next page.)

Torque ratio

It has already been stated that the gearbox multiplies engine torque; thus the relationship between gearbox input torque and output torque can be expressed as a ratio.

Torque ratio   =

It does not follow that the torque ratio will be the same as the gear ratio, this is because some of the input torque is used to overcome friction in the gear assembly.

Efficiency

The efficiency of a gear train is an indication of the friction present. This must therefore be taken into account when calculating the torque ratio.

(a) What would be the efficiency of a gear train if the gear ratio and torque ratio were the same?

....

(b) How is efficiency calculated?

....

....

(c) State the effects of heavy oil in a gearbox compared with a thin oil on the following:

(1) Efficiency ....

(2) Torque ratio ....

(3) Gear ratio ....

(d) In second gear the gearbox input torque is 250 N m and the output torque is 400 N m. Calculate the efficiency of the gearbox if the second gear ratio is 2 : 1

Torque ratio =

Efficiency =

(e) State the gear ratios for a modern vehicle which is fitted with a four speed and reverse gearbox.

Vehicle make and model .... Year ....

Ratios

1st gear .... 2nd gear .... 3rd gear ....

4th gear .... reverse ....

CALCULATIONS: GEAR RATIO

(a) The input gear of a pair of gearwheels has 12 teeth and the output gear has 30 teeth; the gear ratio is:

(a) 2.5 : 1   (b) 2.75 : 1   (c) 3 : 1   (d) 18 : 1

Answer ....

(b) The constant-mesh gearwheels in a sliding-mesh gearbox have 14 and 28 teeth respectively. Calculate the gear ratio for this pair of gears and the speed of the layshaft when the primary shaft speed is 60 rev/s.

....

....

....

....

(c) With a vehicle in second gear and the engine speed at 70 rev/s, calculate the second gear ratio if the propeller shaft speed is 28 rev/s.

....

....

....

(d) A gearbox has a ratio of 3 : 1 in second gear. When the primary shaft speed is 50 rev/s calculate the speed of the mainshaft.

....

....

....

....

(e) Complete the table below.

Gear ratio	4.5 : 1	3 : 1	
Input shaft speed (rev/s)	54		49.5
Output shaft speed (rev/s)		15	18

CALCULATIONS: TORQUE RATIO

(a) Calculate the efficiency of a gearbox given the following data: third gear ratio 1.5 : 1, engine torque 250 N m, propeller shaft torque 350 N m.

....

....

....

....

(b) If the gear ratio of a simple gear train is 3 : 1 and the efficiency is 90%, calculate the output torque and torque ratio when the input torque is 200 N m.

....

....

....

....

(c) In first gear the gear ratio is 4 : 1 and the efficiency of a gearbox is 85%, calculate the torque ratio when the torque transmitted by the engine is 350 N m.

....

....

....

....

(d) Complete the table below.

Gear ratio	2.5 : 1	4 : 1	
Torque ratio	2.5 : 1		6 : 1
Efficiency		95%	93.75%

Compound gear trains

In the motor-vehicle gearbox, to obtain a gear reduction and to facilitate gear-changing, we use two pairs of gears arranged as shown below. This is known as a 'compound' gear train.

To calculate the gear ratio of such an arrangement, we multiply the ratio of one pair of gears by the ratio of the other pair. For example:

$$\text{Gear ratio} = \frac{\text{number of teeth on follower}}{\text{number of teeth on driver}} \times \frac{\text{number of teeth on follower}}{\text{number of teeth on driver}}$$

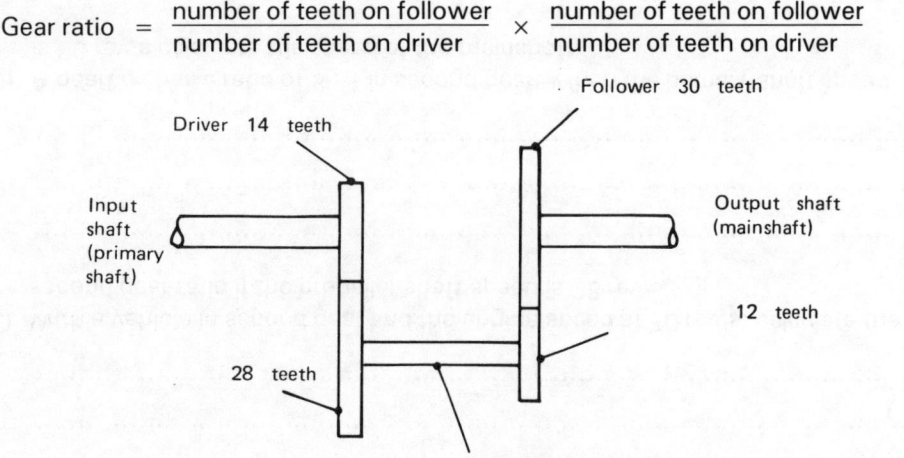

Driver 14 teeth
Follower 30 teeth
Input shaft (primary shaft)
Output shaft (mainshaft)
12 teeth
28 teeth
Intermediate shaft (layshaft)

Complete the labelling on the drawing at C above and calculate the gear ratio.

Gear ratio = ―――――― ―――――― =

State the relationship between each pair of gearwheels.

Input driver	Input follower	Output driver	Output follower	Gear ratio
15	25		30	
	30	15	35	

Problems

1. In a certain gearbox the constant-mesh wheels have 14 and 30 teeth respectively. The first gearwheel on the mainshaft has 28 teeth and the first gearwheel on the layshaft has 16 teeth. Calculate the first gear ratio.

....

....

....

....

2. In a three-speed gearbox the constant-mesh gears have 24 and 36 teeth respectively. The first gearwheel on the mainshaft has 45 teeth and the meshing pinion on the layshaft has 15 teeth. Calculate:

(a) gear ratio in first gear

(b) rev/s of propeller shaft, when engine speed is 50 rev/s.

....

....

....

Overall transmission ratio

To obtain the overall transmission ratio for a vehicle the gearbox ratio is multiplied by the final-drive ratio, the final drive being simply another pair of gears.

Complete the following table to demonstrate this fact.

Gearbox ratio	Final-drive ratio	Overall ratio
2.5 : 1	5 : 1	
7.1 : 1	4 : 1	
1.0 : 1	4.35 : 1	

Problems

1. The final-drive crown wheel on a vehicle has 30 teeth and the pinion has 6 teeth. If the gearbox ratio in first gear is 4 to 1, the overall transmission ratio in first gear would be:

   (a) 20 : 1   (c) 1 : 1

   (b) 9 : 1   (d) 10 : 1

   Ans. (   )

2. In a four-speed gearbox the constant-mesh pinions have 20 and 35 teeth respectively. The second gear on the mainshaft has 30 teeth, and the meshing layshaft gear has 25 teeth. If the rear-axle ratio is 5.5 to 1, calculate:

   (a) the overall gear ratio in second gear

   (b) the propeller shaft speed when the engine speed is 80 rev/s.

   ....
   ....
   ....
   ....
   ....

3. The constant-mesh gears in a gearbox have 16 and 28 teeth respectively. If the second-gear layshaft wheel has 18 teeth, calculate the number of teeth on the second-gear mainshaft wheel.

   ....
   ....
   ....
   ....

4. Complete the table below, given that efficiency =

   $$\frac{\text{Torque ratio}}{\text{Gear ratio}} = \frac{100}{1}$$

Gear ratio	3 : 1	2.5 : 1	
Torque ratio	2.8 : 1		5.4 : 1
Efficiency		97%	91.5%

5. In a four-speed gearbox the constant-mesh gears have 22 and 40 teeth respectively and the first gearwheel on the mainshaft has 42 teeth. Calculate the torque ratio for the gearbox if its efficiency in first gear is 95%.

6. When accelerating in second gear the engine torque for a vehicle is 300 N m and the propeller-shaft torque is 648 N m. If the efficiency of the gearbox in second gear is 96%, calculate the gear and torque ratios.

INVESTIGATION

To determine the gearbox and final-drive ratios of a vehicle.

1. Jack up *one* driving wheel just clear of the ground and remove the engine sparking plugs.

2. Put a chalk mark on the tyre and one on the floor to line up with it. Similarly, mark the crankshaft pulley and timing case (or use ignition timing marks).

3. Engage top gear and count the number of rotations of the engine for one revolution of the jacked-up wheel.

4. Assuming a top gear ratio of 1 to 1, the rear-axle ratio will be:

$$\textbf{axle ratio} = \frac{\text{engine revs}}{2}$$

$$\therefore \textbf{axle ratio} = \frac{}{2} =$$

State the reason why it is necessary to divide by two.

....

....

....

....

State the reason why only one rear wheel is jacked up.

....

....

....

....

....

....

5. Engage first gear and repeat the operation.

$$\text{Overall transmission ratio in first gear} = \frac{\text{engine revs}}{2}$$

$$\therefore \text{ overall transmission ratio in first gear} = \frac{}{2} =$$

$$\textbf{1st gear ratio} = \frac{\text{overall ratio}}{\text{axle ratio}}$$

$$\therefore \textbf{1st gear ratio} =$$

6. Repeat for second gear:

....

....

....

....

7. Repeat for third gear:

....

....

....

....

....

8. Repeat for reverse gear:

....

....

....

....

Chapter 3

Automatic Transmission Systems

ELEMENTS 21/39/42   **UNITS 23/45/46**

AUTOMATIC TRANSMISSION

An automatic transmission system fulfils exactly the same requirements as a manual transmission in that it:

(a) Multiplies engine torque to suit varying load and speed requirements.

(b) ....

(c) ....

In addition to these functions the automatic transmission provides automatic gear changing, that is, the gearbox ratios are selected automatically to meet the speed and load requirement of the vehicle.

What is the difference between FULLY automatic transmission and SEMI-automatic transmission?

....

....

....

....

Most fully automatic transmissions, however, have a manual override facility with which the driver can dictate when gear selections are made. Another feature of modern automatic transmission is a 'mode' selector − for example, car: sport or urban; PSV: Economy or Power drive programmes. These settings affect gear selection according to the way in which the vehicle is being operated. Explain the difference between FIXED RATIO and STEPLESS transmission.

....

....

....

....

....

Show on the drawing opposite (by shading) the location of the automatic transmission unit and state a make and model using it.

Layouts

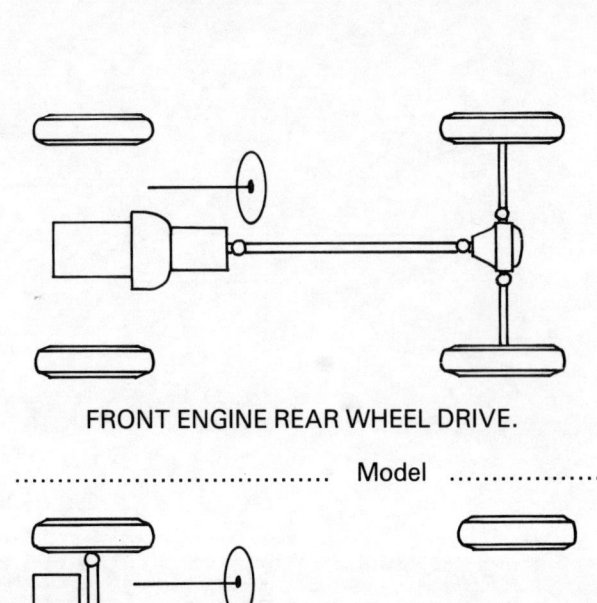

FRONT ENGINE REAR WHEEL DRIVE.

Typical make .... Model ....

FRONT ENGINE FRONT WHEEL DRIVE.

Typical make .... Model ....

REAR ENGINE REAR WHEEL DRIVE.

Typical make .... Model ....

FLUID FLYWHEEL

A fluid flywheel is a hydraulic coupling which is used as an automatic clutch in the transmission system of a vehicle. The unit consists of two main elements, an impeller or driving member and a rotor or driven member. The unit is almost completely filled with fluid. As the engine, and hence the impeller, rotate, the fluid begins to circulate; this transfers torque to the rotor and consequently the gearbox input shaft.

Name the type of gearbox used with this type of coupling:

....

....

Add arrows to the drawing opposite to show the direction of fluid circulation.

State the reason why this type of coupling is generally not used with a normal gearbox:

....

....

State two advantages and two disadvantages of a fluid flywheel as opposed to a friction clutch:

Advantages

1. ....

....

....

2. ....

....

....

Disadvantages

1. ....

....

....

2. ....

....

....

Fluid flywheel assembly

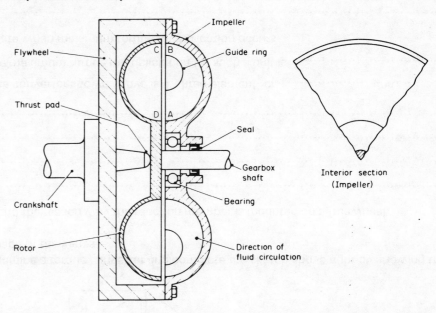

Labels: Flywheel, Impeller, Guide ring, Thrust pad, Seal, Gearbox shaft, Crankshaft, Bearing, Rotor, Direction of fluid circulation

Interior section (Impeller)

Examine the interior of a fluid flywheel assembly and complete the section of the impeller, shown on the right above, by showing the arrangement of the vanes.

Why does the fluid flow from:

1. A to B? ....

....

2. C to D? ....

....

In what other direction is the fluid moving?

....

Describe how the fluid forces the rotor round.

....

TORQUE CONVERTER

The function and action of a torque converter is somewhat similar to that of a fluid flywheel, but with (in its simplest form) the addition of another fixed, bladed member. The advantage of this arrangement is that when 'slip' is taking place a torque multiplication is obtained.

State the type of gearbox normally used in conjunction with a torque converter.

....

A simple line-diagram representing a single-stage three-element torque converter coupling is shown opposite. Complete the labelling on the drawing and add arrows to indicate the path of oil flow. One significant constructional feature of a torque converter is the shape of the vanes in both the impeller and turbine. The greater the change in fluid direction after striking the turbine vanes, the greater will be the force on the turbine. This change in fluid direction is achieved by curving the vanes.

State the reason why it is necessary for the fluid to pass through a reaction member before going back into the impeller.

....

....

....

....

....

....

....

Torque converter

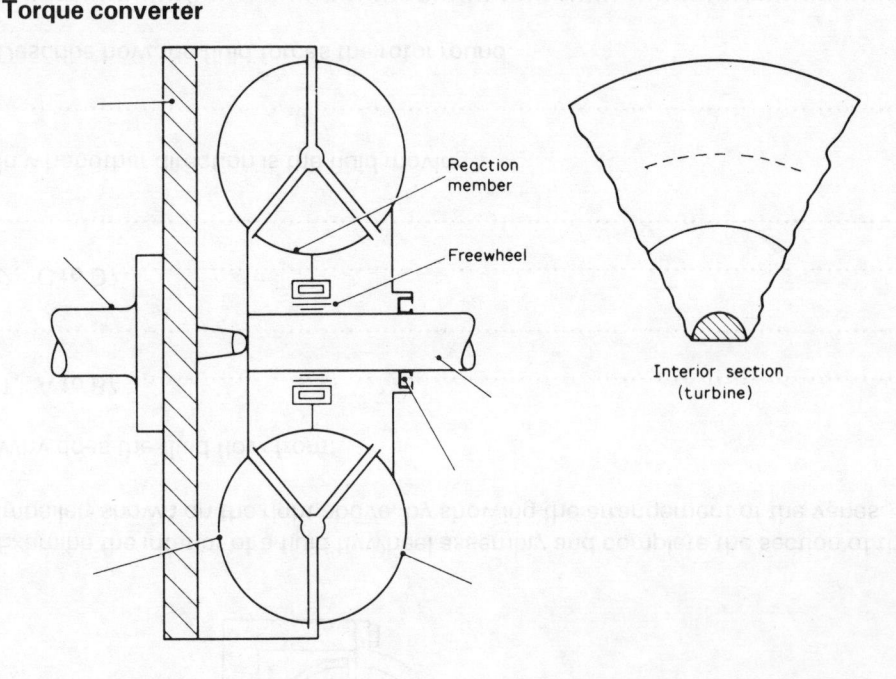

Reaction member

Freewheel

Interior section (turbine)

Examine a torque converter and complete the sketch above right by showing the shape of the vanes.

State the reason why the reaction member is mounted on a freewheel.

....

....

....

....

The converter shown above is a 'three-element' or .... unit,

the maximum torque multipliclation for such a unit is ....

State when maximum torque multiplication occurs.

....

....

COUPLING CHARACTERISTICS

The graph at (A) shows the efficiency output curve for a constant input speed. For the fluid flywheel, complete the graph at (B) to show the curve for a torque converter.

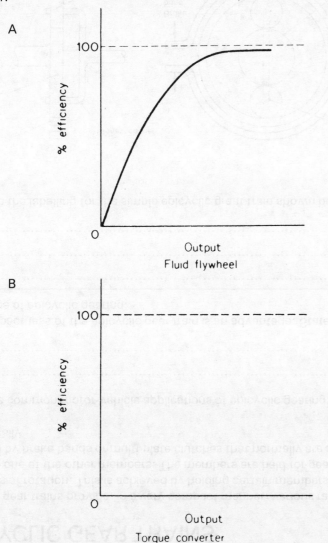

A

Output
Fluid flywheel

B

Output
Torque converter

Considering the formula KINETIC ENERGY = $\frac{1}{2}mv^2$, state the relationship between fluid speed and kinetic energy.

....
....
....
....

The velocity and hence the kinetic energy of the fluid is increased as it moves to the outer radius of the impeller. State how the kinetic energy is converted to a force which produces a driving torque in the turbine or output shaft.

....
....
....
....
....

Some energy is lost because of conversion to heat energy, that is, the fluid is heated as it is made to work on the turbine. Under which operating condition is maximum heat generated?

....
....
....
....
....
....

EPICYCLIC GEAR TRAINS

21.3

Epicyclic gear trains provide, in a very compact manner, various ratios and directions of rotation. This is achieved by holding certain members and applying power to one of the other members. The members are held for gear engagement purposes by brake bands or multi-plate clutches that normally are actuated hydraulically.

List some common motor-vehicle applications of epicyclic gearing:

....

....

The compactness of the epicyclic gear train is an advantage. State one other advantage of epicyclic gearing:

....

....

Complete the labelling for the simple epicyclic great train shown below.

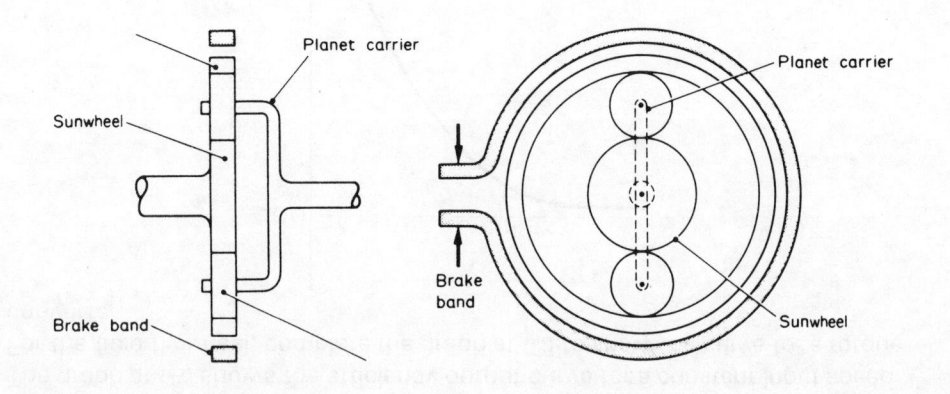

INVESTIGATION

Examine a simple epicyclic gear train and study the three combinations shown below.

Describe alongside each drawing the relationship between input and output in each case. Add arrows to the drawings to indicate the direction of rotation of the wheels.

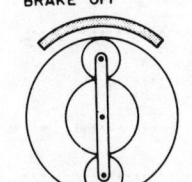

Input sunwheel – output planet carrier

....

....

....

....

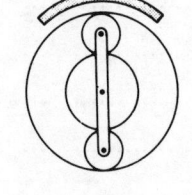

Input sunwheel – output annulus – planet carrier locked

....

....

....

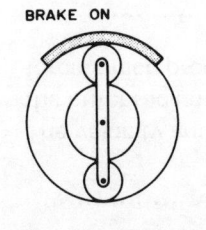

Input planet carrier – output sunwheel

....

....

....

Describe how it is possible to increase the number of gear ratios available using epicyclic gearing.

....

....

....

Compound epicyclic gear trains

This is a more complex and sophisticated type of epicyclic arrangement than the simple epicyclic gearing mentioned on the previous page. It has the advantages of being able to obtain a greater number of forward ratios, and reverse.

A 'two-element' epicyclic gear train, as used in some automatic gearboxes, is shown opposite; this gear set will provide three forward gears and one reverse gear. One important feature to appreciate is the fact that the planet carrier is mounted on a freewheel or one-way clutch; this will only permit the carrier to revolve clockwise (as seen from the drawing). Examine such a gear set and, in the spaces provided describe its operation in each gear.

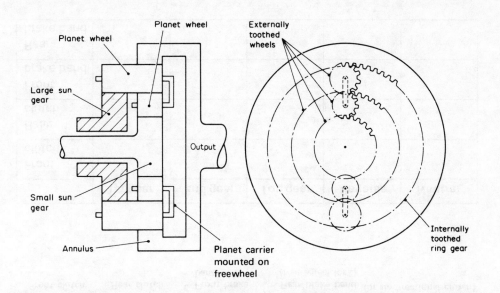

....

....

....

1st gear, input to small sun gear

....

....

....

....

....

2nd gear, input to small sun gear – large sun gear locked

....

....

....

....

....

Top gear, small sun gear and large sun gear locked together – input through both

....

....

....

....

....

Reverse gear, input to large sun gear – small sun gear free – freewheel locked

....

....

....

....

....

AUTOMATIC GEARBOXES

The drawing opposite represents the complete mechanical system in an automatic gearbox. The various ratios are obtained through a compound epicylic gear set such as outlined on the previous page.

Although the automatic gearbox arrangement shown opposite is an early model, it is a relatively simple example and thus ideal for the purpose of gaining an understanding of automatic gearbox operation.

Study the control system for the gear set and indicate, on the table below the drawing, which clutches and brake bands are operative in each gear.

State:

1. The conditions under which the freewheel is operating, and
2. The effect of this action.

....

....

....

State the effect of the application of the rear brake band in first and second gears:

....

....

State the reason why the rear brake band is applied in reverse gear:

....

....

....

....

....

....

....

Mechanical system fully automatic gearbox

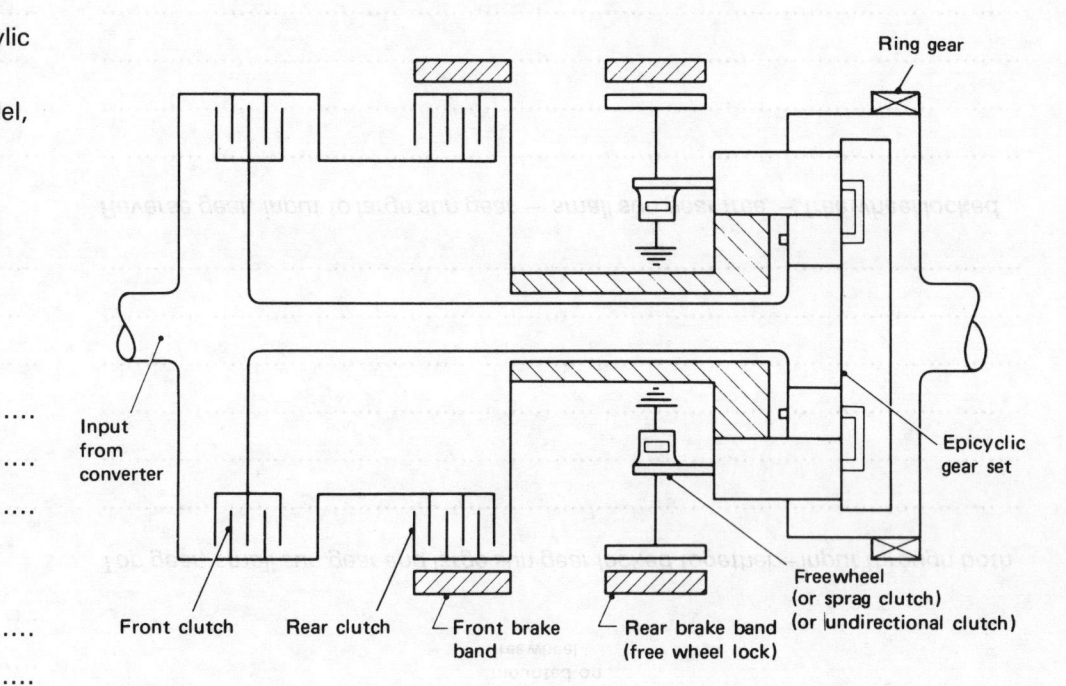

Input from converter

Ring gear

Epicyclic gear set

Freewheel (or sprag clutch) (or unidirectional clutch)

Front clutch   Rear clutch   Front brake band   Rear brake band (free wheel lock)

	1st gear	2nd gear	Top gear	Reverse	Neutral
Front clutch					
Rear clutch					
Front brake band					
Rear brake band					

Automatic gearbox hydraulic system

The hydraulic system in the automatic gearbox serves three basic purposes. It:

1. maintains a pressurised supply of fluid to the torque converter;
2. provides gearbox lubrication;
3. automatically controls the application and release of the brake bands and multi-plate clutches.

The brake bands and clutches are actuated by hydraulic servo units.

Complete the drawing below to show the application of the brake band by a hydraulic servo.

The hydraulic pressure required to operate clutch and brake-band servos, for lubrication and for torque converter operation, is provided by a pump driven by the impeller shaft of the torque converter.

Five basic control units are responsible for the action of the hydraulic system; these are:

1. .... 2. ....

3. .... 4. ....

5. ....

These, together with the pump, are represented in the simple block diagram below; label the valves and complete the diagram to show how the units are interconnected within the hydraulic circuit.

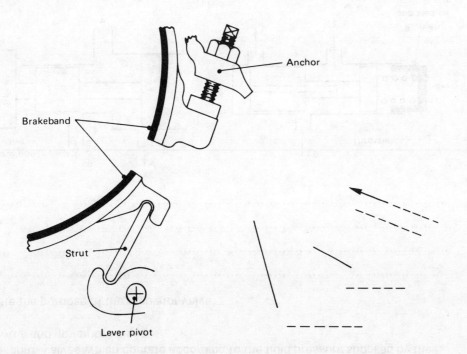

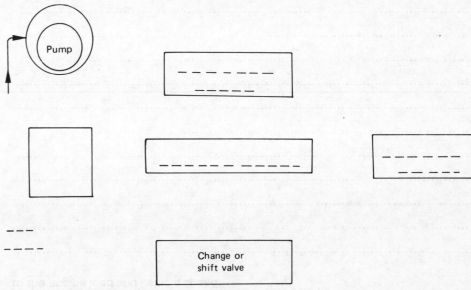

Change or shift valve

State the reason why the fluid level changes when the selector lever is moved:

....

Principle of operation (automatic gearchange)

Automatic application and release of the clutch or brake servos are controlled by the 'shift' valves which operate according to the fluid pressure supplied by the throttle and governor valves.

State the purpose of the regulator valve:

....

....

....

....

The simple drawing on the left shows the shift valve directing line pressure to a clutch or brake servo.

Describe how the throttle and governor valves operate the shift valve according to the engine load and vehicle speed:

....

....

....

....

....

....

....

....

....

....

....

....

....

....

....

....

....

....

....

....

Governor valve

Weight

LP

V

LP

Throttle valve

Plunger-operated by throttle cam

GP

LP

Shift valve

TP

V

LP—Line pressure

V—Vent pressure

GP—Governor pressure

TP—Throttle pressure

Servo applied

Servo exhaust

Gear selection (automatic transmission)

A selector lever and quadrant showing the various operating positions for a three-speed automatic transmission is shown below.

State the purpose of each gear-selector position:

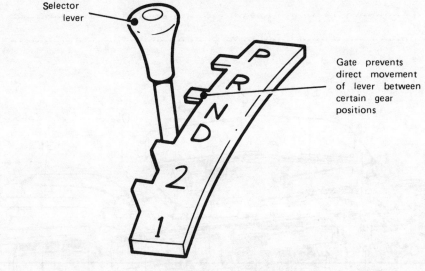

Selector lever

Gate prevents direct movement of lever between certain gear positions

P. ....
....
....
R. ....
N. ....
D. ....
....
....

2. ....
....
....
....
....
....

1. ....
....
....

Starter inhibitor

For safety reasons the starter can only be operated in selector positions P and N. This is achieved by an inhibitor switch on the gearbox which is operated by the manual selector mechanism. The four-terminal switch is operative in two positions for starter, or reverse-light operation.

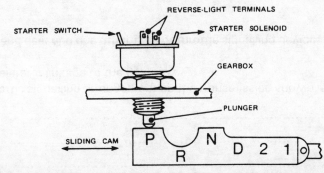

REVERSE-LIGHT TERMINALS
STARTER SWITCH
STARTER SOLENOID
GEARBOX
PLUNGER
SLIDING CAM
P R N D 2 1

State the precautions needed when towing a vehicle with automatic transmission when:

(a) engine is defective ....
....

(b) gearbox is defective ....
....

A complete automatic transmission is shown, ZF (BMW), below. Name the numbered parts.

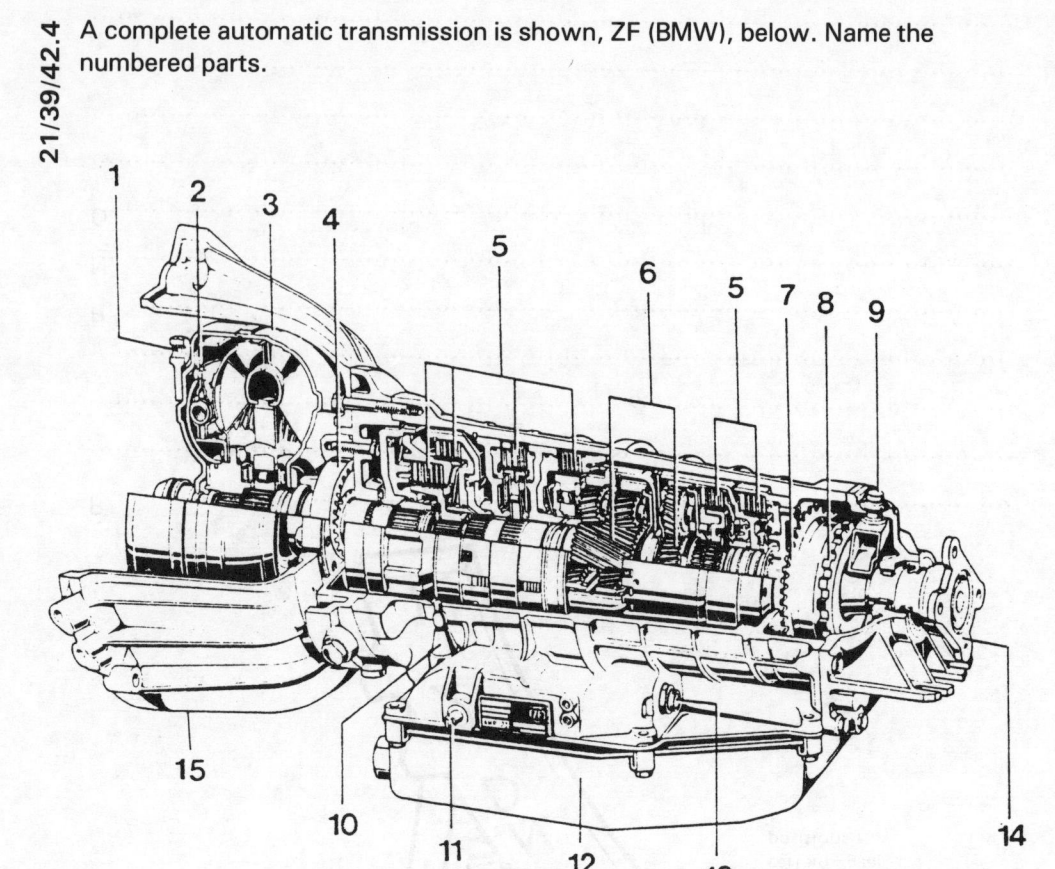

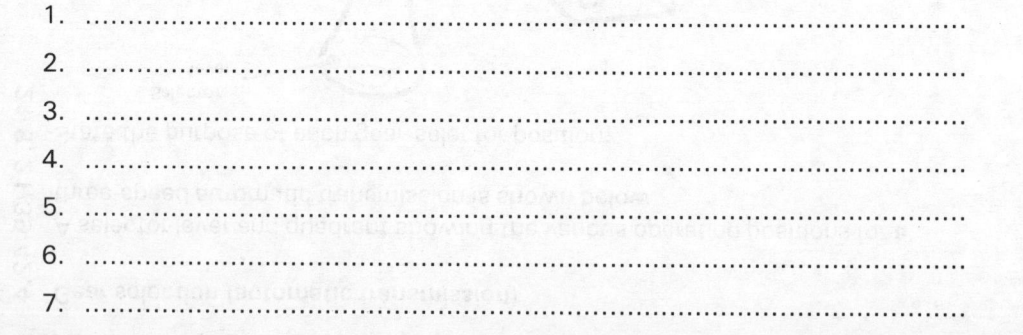

1. ....
2. ....
3. ....
4. ....
5. ....
6. ....
7. ....

8. ....
9. ....
10. ....
11. ....
12. ....
13. ....
14. ....
15. ....

The throttle signalling arrangement on the transmission shown is a cable from the accelerator linkage to the gearbox.

Sketch and describe one other type of throttle signalling device.

VALVE BLOCK

CONTROL SYSTEM

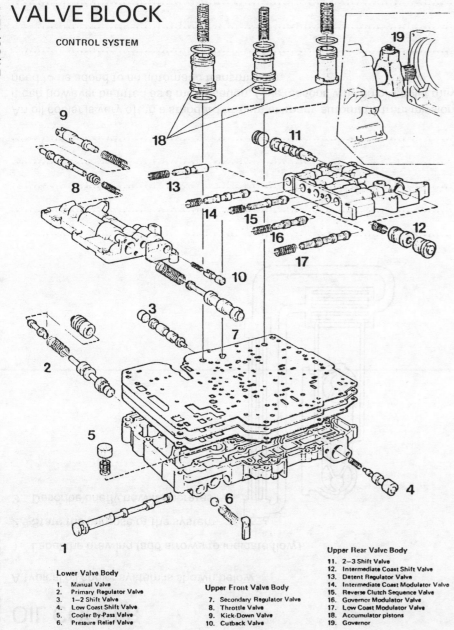

Lower Valve Body

1. Manual Valve
2. Primary Regulator Valve
3. 1–2 Shift Valve
4. Low Coast Shift Valve
5. Cooler By-Pass Valve
6. Pressure Relief Valve

Upper Front Valve Body

7. Secondary Regulator Valve
8. Throttle Valve
9. Kick-Down Valve
10. Cutback Valve

Upper Rear Valve Body

11. 2–3 Shift Valve
12. Intermediate Coast Shift Valve
13. Detent Regulator Valve
14. Intermediate Coast Modulator Valve
15. Reverse Clutch Sequence Valve
16. Governor Modulator Valve
17. Low Coast Modulator Valve
18. Accumulator pistons
19. Governor

A typical hydraulic control system, Borg Warner (VOLVO), is shown opposite. State the function of the accumulator:

....

....

....

....

....

Name the types of fluid seal used and their particular application in an automatic transmission:

....

....

....

Name, state the purpose and describe the operation of the component shown below:

pressure relief valve

....

....

....

....

OIL COOLER

A typical oil cooler system is shown below.

1. Label the drawing (add arrows to indicate flow).

2. State the purpose of the system.

3. Describe briefly how it operates.

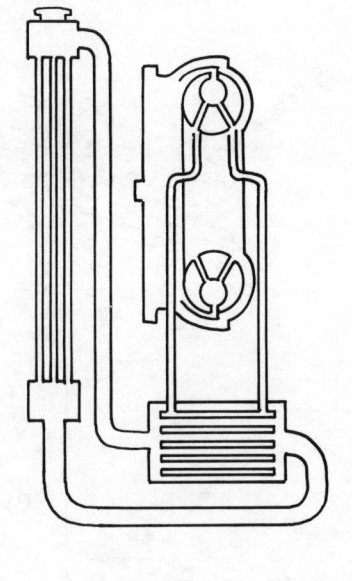

....
....
....
....
....

An oil cooler is very often a standard fitment on many automatic transmissions, it can however be fitted as a modification. Give reasons why an oil cooler may need to be added to an automatic transmission.

....
....
....

ELECTRONIC CONTROL

An electronically controlled automatic gearbox is basically the same as a hydraulically controlled automatic gearbox in that the main components are torque converter, epicyclic gear set, multi-plate clutches and hydraulic servos.

In addition to these main components, electronic shifting entails the use of:

1. ECU 2 ....   3. ....

Complete the simple block diagram for such a system and describe briefly how automatic gear changing is achieved.

SENSORS

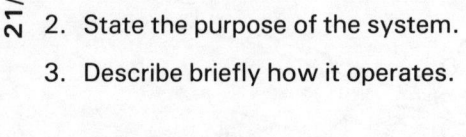

1 vehicle speed ....→
2 ....→
3 ....→
4 ....→   E C U
5 ....→
6 ....→
7 ....→

solenoid / HYDRAULIC CIRCUIT → clutch servos
solenoid

Operation

....
....
....
....
....
....
....

Name one typical vehicle using electronic control.

Make ....   Model ....

CONTINUOUSLY VARIABLE AUTOMATIC TRANSMISSION

Continuously variable or STEPLESS transmission is an alternative to the FIXED RATIO transmission. Unlike conventional automatic transmission, the gear ratios are varied in a smooth, stepless progression to suit driving condition (speed and load).

One transmission arrangement contains two basic elements:

(a) a PLANETARY GEARSET integrated with two wet multi-plate clutches;

(b) a BELT and PULLEY system.

The components named at (a) provide take up from rest and drive to the pulley system.

State the function of (b).

....
....
....
....

A front engine front wheel drive transaxle (FORD) is shown opposite. Complete the labelling on the drawing.

This system does not require a torque converter or a conventional friction clutch. Why is this?

....
....
....

List two vehicles using this (or similar) forms of transmission.

Make .... Model .... Engine size ....

Make .... Model .... Engine size ....

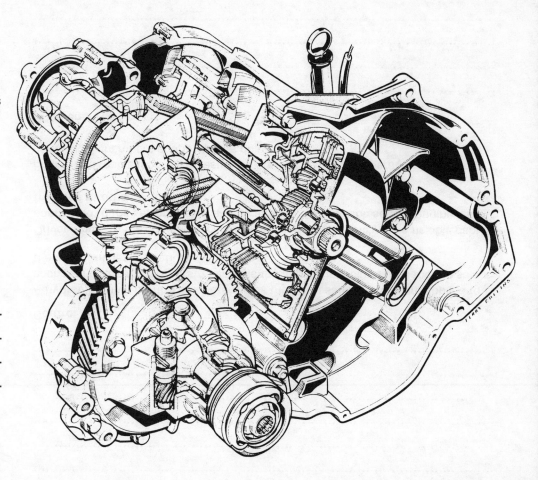

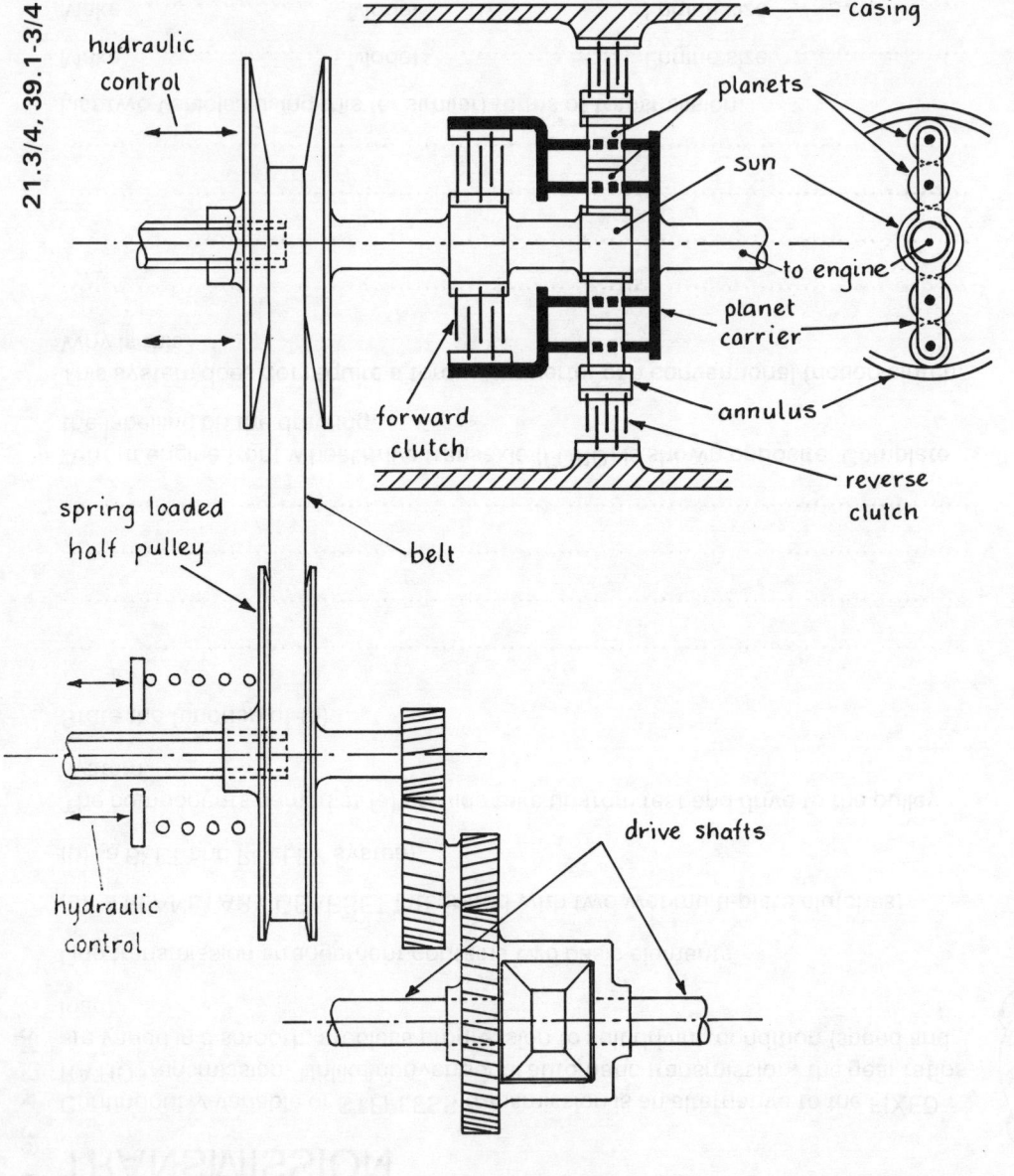

casing

planets

sun

to engine

planet carrier

annulus

hydraulic control

spring loaded half pulley

belt

forward clutch

reverse clutch

hydraulic control

drive shafts

Operation

The forward clutch is gradually engaged by a hydraulic servo, the oil being supplied by an engine driven oil pump via a control unit.

With the forward clutch clamped, the sunwheel and planet carrier are locked together making a fixed drive to the primary driving pulley.

Describe the action of the gearset in reverse.

....

....

....

....

....

....

....

Control

A hydraulic valve-control box determines the oil pressures which are applied to each part of the system: first to engage the appropriate clutch and then to select the optimum gearing ratio.

Driver inputs to the control box are transmitted mechanically via a flexible cable connected to the selector lever and by another cable which senses engine load from the throttle position.

Engine and road speed are measured by two PITOT tubes operating in rotating centrifugal chambers at appropriate points.

What are PITOT tubes?

....

....

....

....

Describe, with the aid of diagrams, the action of the belt and pulley system.

....
....
....
....
....
....
....
....
....
....
....
....

Driving pulley width is controlled by: ....

....

Secondary or driven pulley width is controlled by:

....
....
....

From what material is the belt made?

....
....
....

LOW GEAR   HIGH GEAR

State the advantages of the continuously variable automatic transmissions over the fixed ratio automatic transmission.

....
....
....
....

67

BAND OPERATED AUTOMATIC GEARBOX

This system consists of a fluid flywheel in series with a four- or five-speed epicyclic gear set. Brake bands, operated by hydraulic servos, are applied and released to control the power flow and ratio selection in the gear set.

Brake band operation

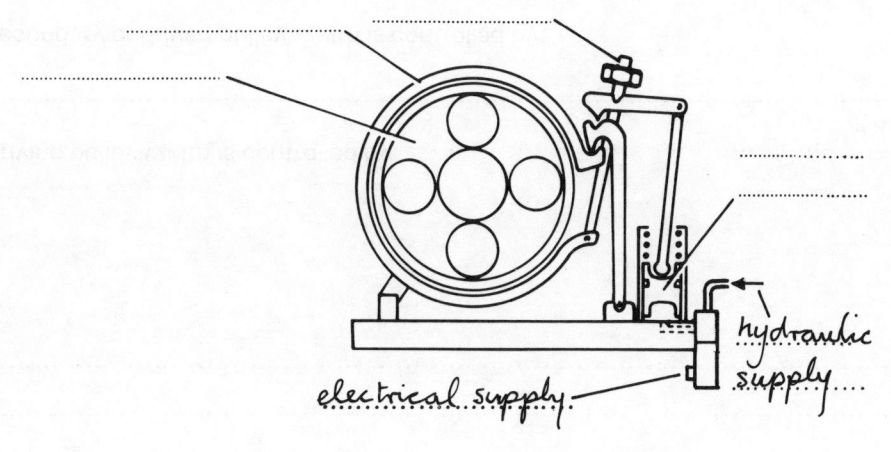

electrical supply.

hydraulic supply

Complete the labelling on the drawing and briefly describe the action of the system during brake band application.

....

....

....

....

....

....

The drawing below illustrates the layout of the control system for a fully automatic (HYDRACYCLIC) transmission popular on PSVs. Describe the operation of the system.

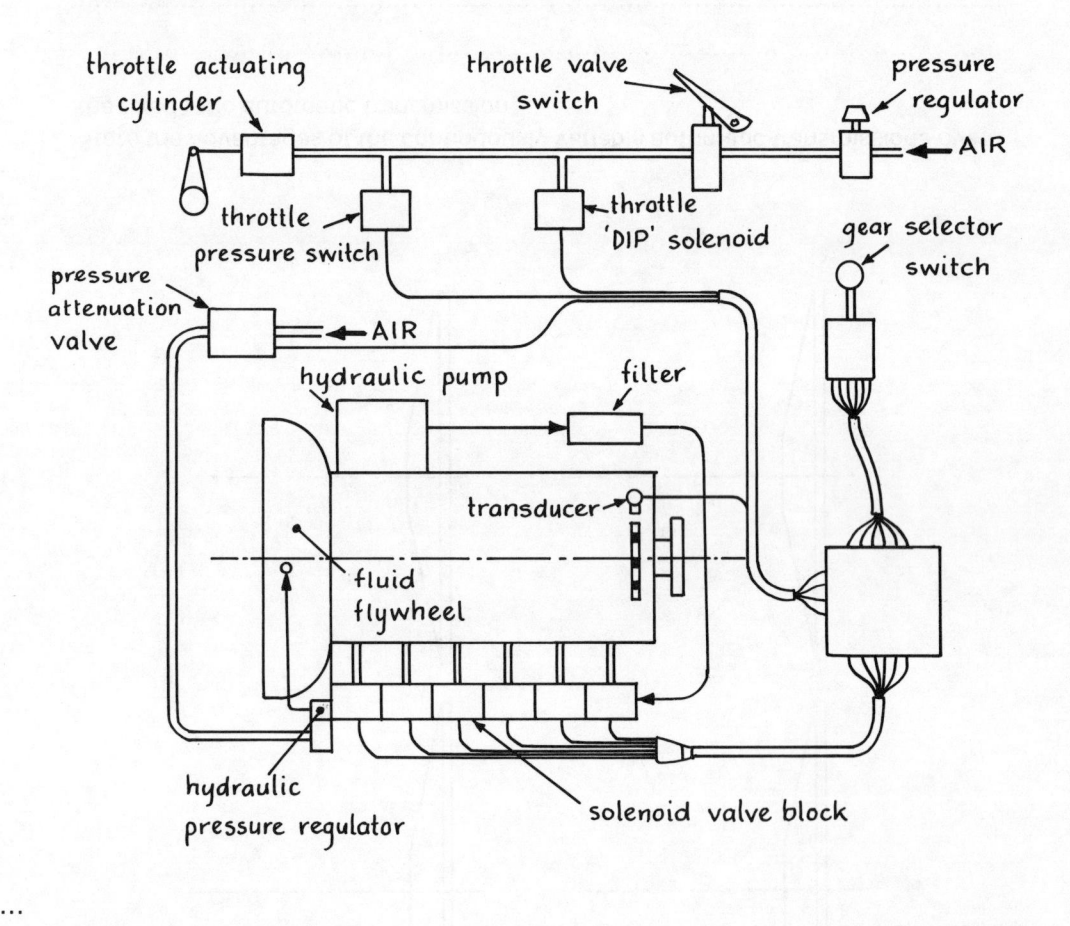

throttle actuating cylinder

throttle valve switch

pressure regulator

AIR

throttle pressure switch

throttle 'DIP' solenoid

gear selector switch

pressure attenuation valve

AIR

hydraulic pump

filter

transducer

fluid flywheel

hydraulic pressure regulator

solenoid valve block

Operation

....

....

....

....

Operation (continued)

....
....
....
....
....
....
....
....
....
....
....
....
....
....

The system usually incorporates a 'friction retarder', which comes into operation when the brakes are applied, and a 'lock up' clutch. What is the purpose of a lock up clutch?

....
....
....

An alternative to the fully automatic system is shown opposite. This is a SEMI-AUTOMATIC system which utilises the same gearbox. The brake band servos are applied hydraulically as in the fully automatic system; in earlier gearboxes compressed air was used.

SEMI-AUTOMATIC HYDRACYCLIC GEARBOX

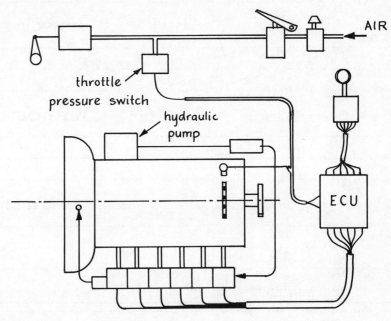

Operation

As the driver selects the gears, the electrical solenoids are energised to actuate the hydraulic valves. What is the purpose of the ECU (TRANSLATOR), Throttle Pressure Switch and Transducer in this system?

....
....
....
....
....
....

COMPUTER AIDED GEARCHANGING (CAG)

An optional transmission system for hgvs (SCANIA) is CAG. The physical action of gear changing is performed by air-powered actuating cylinders. It is a form of PRE-SELECTOR transmission.

Main components

Name and state the function of the numbered components.

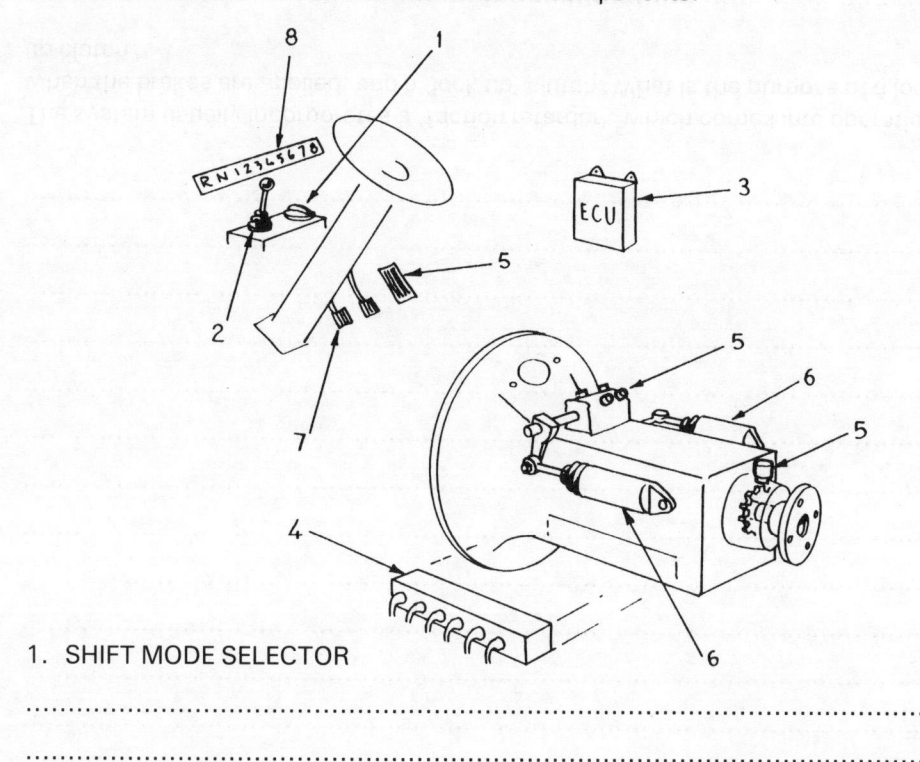

1. SHIFT MODE SELECTOR

....

....

2. SHIFT STALK

....

....

....

....

3. COMPUTER

....

....

....

....

4. SOLENOID VALVES

....

....

....

....

5. SENSORS

....

....

....

....

....

6. ACTUATING-CYLINDERS

....

....

....

7. CLUTCH PEDAL

....

....

8. GEAR DISPLAY

....

....

TESTING AND FAULT DIAGNOSIS

One method of testing a torque converter coupling is to carry out a STALL TEST on the vehicle. Outline this procedure, include any safety measures adopted during the test:

....
....
....
....
....
....
....
....
....
....
....

For satisfactory operation the engine stall speed should be approximately .... revs/min. Indicate the possible faults for the results given below.

Engine speed 300 revs below specified stall speed:

....

Engine speed 600 revs below specified stall speed:

....

Stall speed too high:

....
....
....
....

A road or chassis dynamometer test is another way in which torque converter faults are diagnosed; the road test will also confirm faults suspected during a stall test.

State how the road test would indicate:

1. Slipping stator

....
....
....

2. Seized or locked stator

....
....
....

If the converter pressure is low 'cavitation' will occur, resulting in slippage, noise and vibration. What are the effects if converter pressure is too high?

....
....
....

With the faults at 1 and 2 above it is necessary to replace the converter as it is normally a sealed unit. State the safety precautions to be taken when draining the fluid.

....
....
....
....

The fluid coupling requires very little maintenance apart from topping up with fluid; however, leakage could result from a worn seal and worn bearings could allow the faces of the impeller and rotor to come into contact, giving noisy operation.

71

How can the system be protected during use or repair against the following hazards?

1. Overheating

....
....
....

2. Aeration and foaming

....
....
....

3. Contamination of fluid

....
....
....

4. Ingress of dirt and moisture

....
....
....

5. Fluid leakage

....
....
....

6. Mechanical damage

....
....
....

Routine preventive maintenance will improve reliability and efficiency, and maximise the life of a transmission system. List the preventive maintenance checks and tasks associated with automatic transmission.

Check: ....
....
....
....
....
....
....
....
....
....
....
....
....
....
....
....

List the general rules for efficiency and any special precautions to be observed when carrying out maintenance on the transmission.

....
....
....
....
....
....
....
....
....
....

AUTOMATIC TRANSMISSION SYSTEM FAULTS, SYMPTOMS, PROBABLE CAUSES AND CORRECTIVE ACTION

Complete the table in respect of the faults listed.

FAULTS	SYMPTOMS	PROBABLE CAUSES	CORRECTIVE ACTION
Worn multi-plate clutch			
Worn or incorrectly adjusted selector linkage			
Broken throttle signalling device (cable or vac pipe)			
Incorrectly adjusted kickdown cable			
High engine idling speed			
Low line pressure			
High line pressure			
Oil leakage			
Broken parking pawl			

Chapter 4

Drive Line

ELEMENT 22　　　　**UNITS 16/35/36**

DRIVE LINE SHAFTS AND HUBS

Drive line

Identify the drive line arrangements on this page. Name the major parts and give a brief description of each.

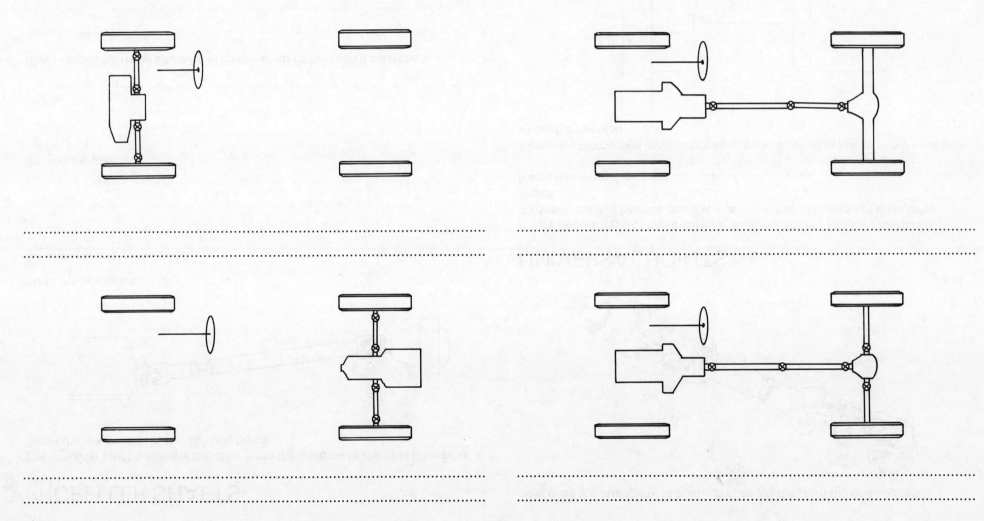

....

....

....

....

22.3 PROPELLER SHAFTS

The propeller shaft transmits the drive from the gearbox to the final drive gear.
Name the major parts in the drawing below.

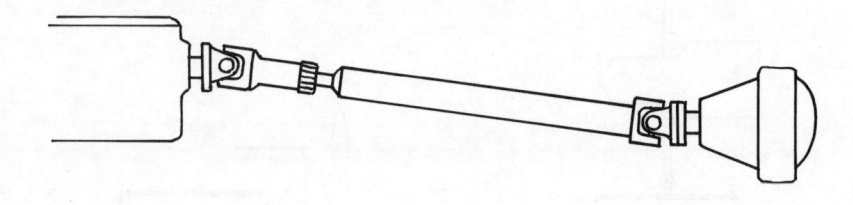

State the purpose of:

(a) the universal joints (UJs)

....
....
....

(b) the slip joint

....
....
....

Give reasons for using a tubular propeller shaft rather than a solid shaft.

....
....
....

What alignment feature is vital during service work on the above shaft?

....
....
....

How does the arrangement below primarily differ from the one opposite?

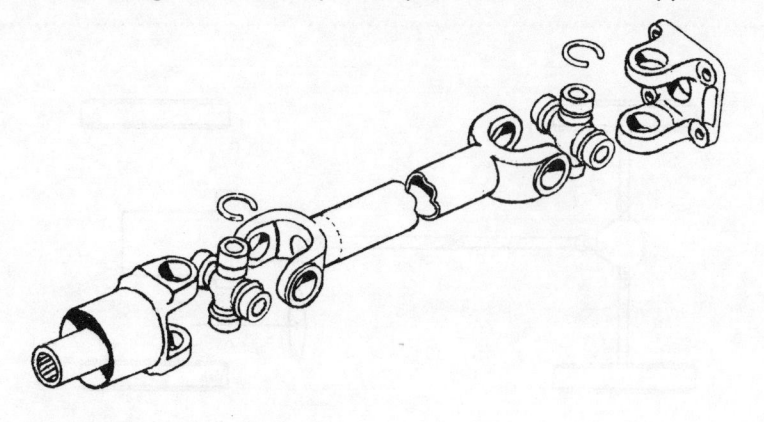

UNIVERSAL JOINTS

A very popular type of universal joint is shown above. The joint is efficient, compact, easy to balance and it will transmit the drive through quite large angles.
Name this type of joint ....

Examine a dismantled joint and complete the drawing below to show a section through a trunnion.

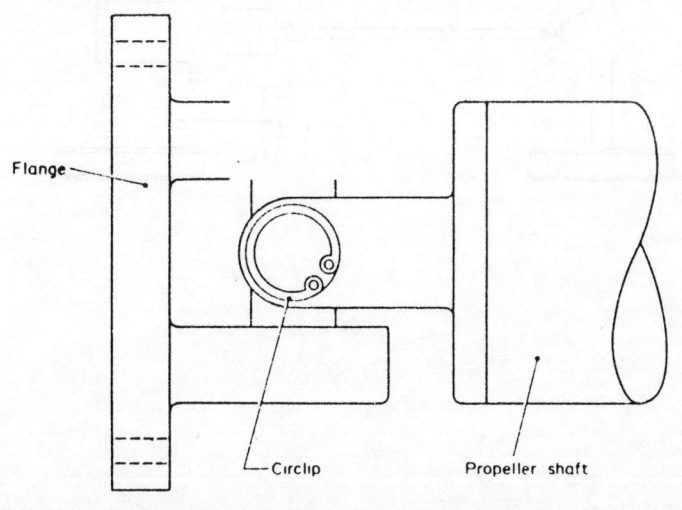

Flange

Circlip

Propeller shaft

The constructional features of two other types of universal joints are shown below; name the two joints and label the drawings.

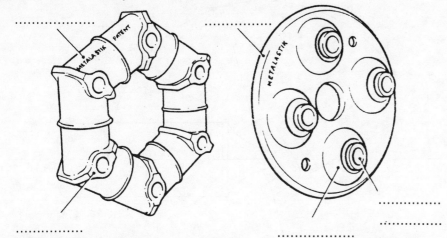

Complete the table below.

Type	Advantages	Disadvantages
Layrub	1. .... 2.	1. .... 2. ....
Hookes		
Doughnut		

Owing to the high rotational speed of a propeller shaft it is essential that the assembly is balanced, otherwise transmission vibration will result. It is now common practice to employ a 'two piece' propeller shaft assembly, with an intermediate or centre bearing providing support.

Examine a vehicle and complete the drawing below to show a two-piece arrangement.

Vehicle make .... Model ....

Label the flexibility mounted centre bearing shown below.

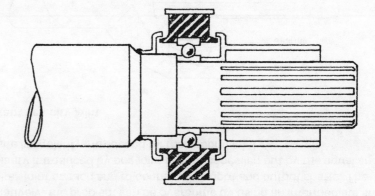

State TWO vehicles which use a propeller shaft centre bearing:

MAKE .... MODEL ....

MAKE .... MODEL ....

Universal joint characteristics

When drive is being transmitted from one shaft to another through a Hookes-type joint, if one shaft swings through an angle its velocity will vary, that is, the shaft will accelerate and decelerate slightly during every 180° of revolution. This is more pronounced if the angle through which the shaft swings is great, for example, the drive to a steered wheel.

State the conditions necessary if a universal joint is to transmit 'constant velocity'.

....
....
....

Hookes-type joint

The drawing below illustrates the problem with a Hookes – type joint; the line B is bisecting the angle between the two shafts. Draw a centre line through the joint driving contacts.

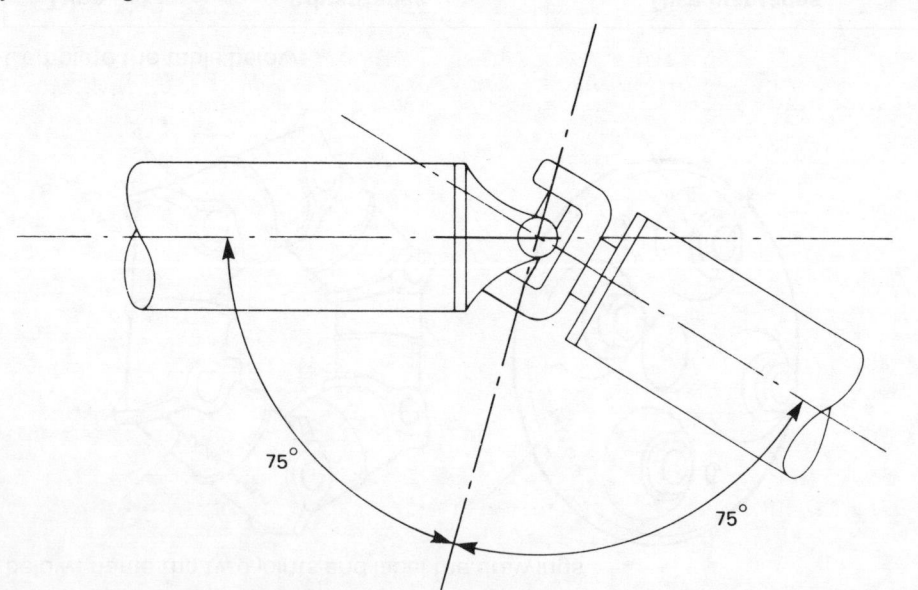

75°

75°

CONSTANT VELOCITY

Constant velocity can be transmitted from one shaft to another through varying angles by a special type of joint known as a 'constant velocity joint'. Alternatively, the problem can be overcome by using an intermediate shaft with a Hookes joint at each end to connect the input and output shafts. The irregularity introduced by one joint is then cancelled out by the equal and opposite irregularity introduced by the second joint.

Constant velocity joint

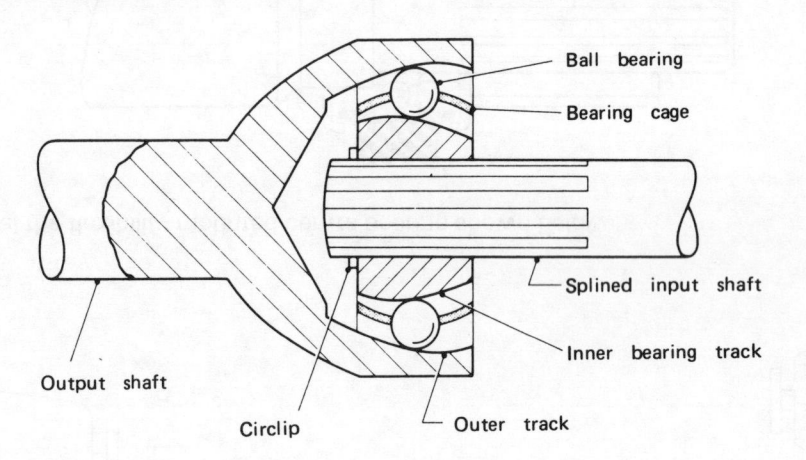

Ball bearing

Bearing cage

Splined input shaft

Inner bearing track

Output shaft

Circlip

Outer track

The joint shown above will transmit constant velocity through varying angles. The driving contacts always lie on a plane bisecting the angle between the input and output shaft. Describe how this is achieved.

....
....
....
....
....
....

Plunge-type constant velocity joint

The plunge-type constant velocity joint is designed to accommodate lateral movement of the drive shaft as the road wheel rises and falls.

Complete the drawing below to show a plunge-type joint.

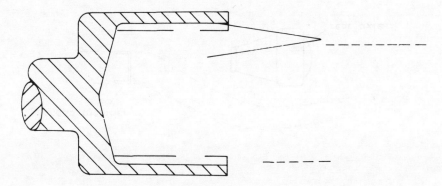

To achieve constant velocity through an intermediate shaft and two universal joints, the intermediate shaft must be arranged so that it makes equal angles with the first and third shafts and the joint yokes on the intermediate shaft must be parallel to one another.

Make a drawing below to show such an arrangement.

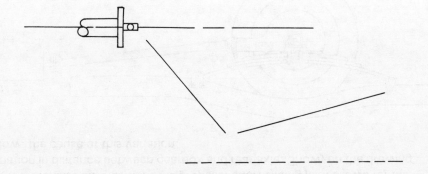

Hotchkiss drive

In this arrangement a 'live' rear axle is mounted on semi-elliptic leaf springs which act as both suspension springs and axle-locating linkage.

Complete the drawing below to show the Hotchkiss layout and indicate on the drawing how the tractive effort, produced at the point where the tyre contacts the road surface, results in a driving thrust on the vehicle frame.

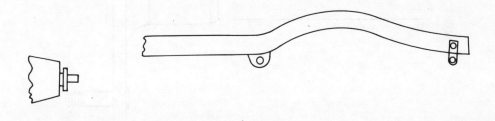

In the Hotchkiss system the driving thrust is transmitted from the wheel and axle casing to the chassis via:

....

....

....

Drive torque reaction, which tries to rotate the axle casing in the opposite direction to the road wheels, is resisted by:

....

....

....

....

....

In the Hotchkiss arrangement the propeller shaft sliding joint allows for the variation in distance between gearbox and rear axle. Show, on the drawing below, the cause of this variation.

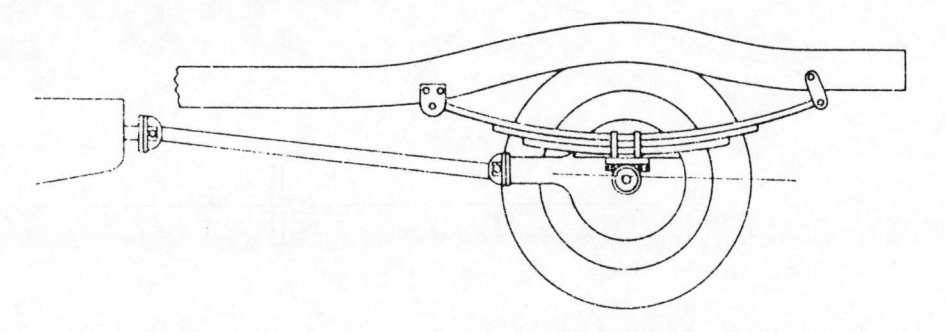

Many cars employ an alternative arrangement to the Hotchkiss type, in which an extension housing (TORQUE TUBE) transmits driving thrust from the final drive casing to the vehicle frame and withstands rear axle drive and brake torque reactions. Examine a vehicle with a torque tube housing and complete the drawing below to show this.

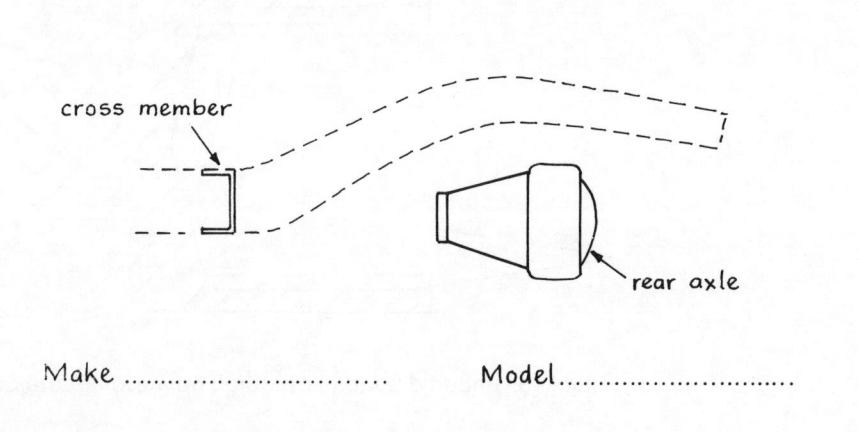

cross member

rear axle

Make ....   Model...........................

Hubs

The number, type and location of bearings employed in a hub assembly depends, to a large extent, on the weight of the vehicle.

Three main classifications of hubs are:

1. *Semi- or non-floating* ....

2. ....

3. ....

Examine semi-, three-quarter, and fully floating rear-hub arrangements (preferably sectioned examples) and complete the drawings on this and the next page to show the main constructional features of each, including any oil or grease seals.

Semi-floating

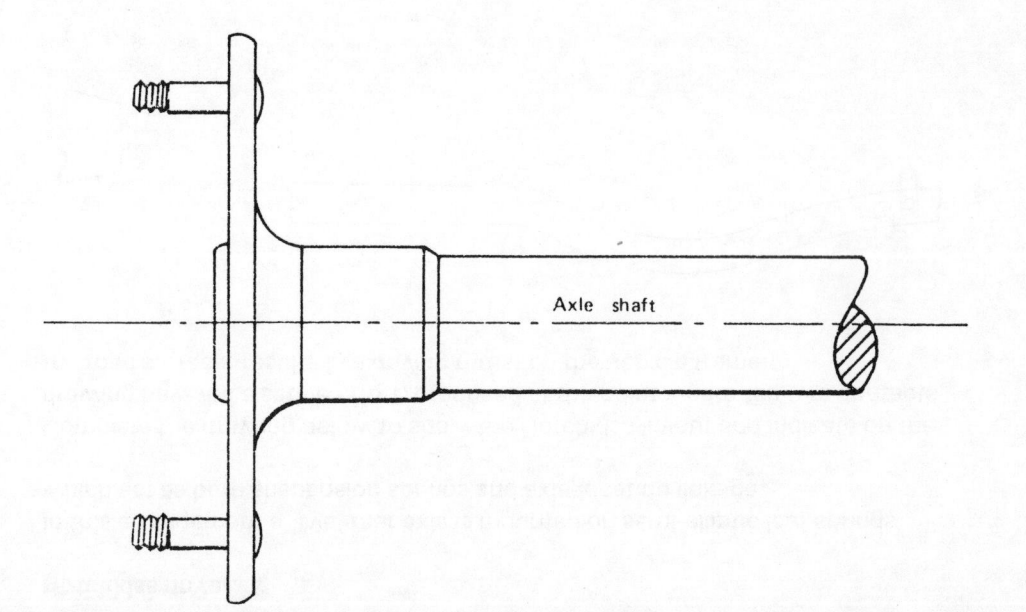

Axle shaft

Three-quarter floating

Fully floating hub

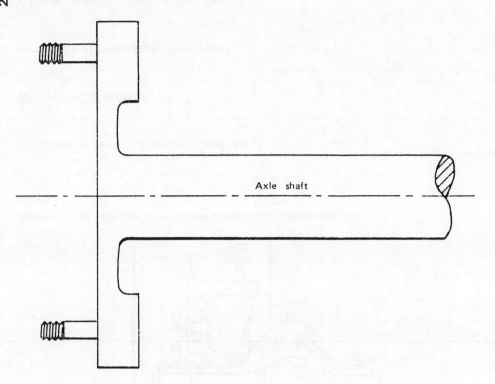

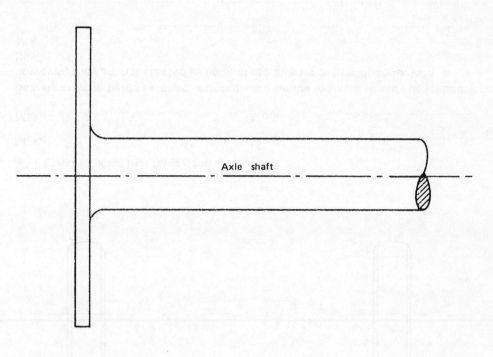

Axle shaft

Axle shaft

Name the vehicle types, using this hub arrangement, and state the main differences between it and the semi-floating type.

....

....

....

....

....

....

Describe how the bearings are adjusted during assembly:

....

....

....

Which types of vehicles use fully floating axles or hubs?

....

What is a 'live axle'?

....

81

STEERED FRONT AXLE

Name the parts of the hub shown below.

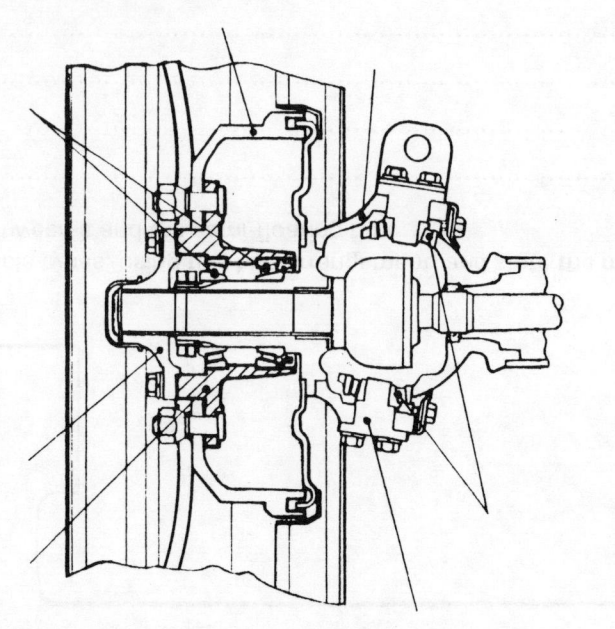

Suggest two vehicles that have a similar hub arrangement:

Make .... Model ....

Make .... Model ....

Is the hub shown for a 'live' or 'dead' axle?

....

Which class of hub is it?

....

What type of bearings are employed?

....

What is the purpose of the drive flange?

....

DEAD AXLE

What is the meaning of the term 'dead axle'?

....

....

Complete the sketch below to show a dead axle.

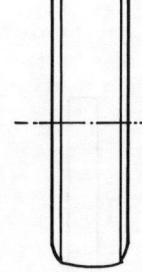

Name two vehicles that have dead axles.

Make .... Model ....

Make .... Model ....

Bearing arrangements in 'dead' axle hubs are similar to those already described, however, they are not classed as semi, three-quarter or fully floating. Why is this?

....

....

....

DRIVELINE SHAFTS AND HUBS: FAULTS, SYMPTOMS, PROBABLE CAUSES AND CORRECTIVE ACTION

Complete the table in respect of the faults listed.

FAULTS	SYMPTOMS	PROBABLE CAUSES	CORRECTIVE ACTION
Worn universal joint trunnions			
Split rubber gaiters			
Worn hub bearings			
Worn centre bearing			
Worn sliding joint oil seal			
Worn constant velocity joint			
Worn driveshaft splines			
Loose shaft flange bolts			
Sheared rubber (doughnut type universal joint)			

PROTECTION DURING USE

How can drive line systems be protected against the ingress of moisture and dirt during use and repair?

....

....

....

....

Describe any special tools (and their care) necessary to carry out routine maintenance and adjustments:

....

....

....

....

....

....

MAINTENANCE

Routine maintenance is essential in order to ensure reliability, maintain efficiency, prolong service life and ensure vehicle safety. List the major maintenance points:

....

....

....

....

....

....

....

....

....

....

List the general rules/precautions to be observed when carrying out routine maintenance adjustments, removal and replacement relative to the drive line, shafts and hubs:

....

....

....

....

....

....

....

....

....

Chapter 5

Final Drive and Differentials

ELEMENTS 23/40/41   **UNITS 16/35/36**

FINAL-DRIVE AND DIFFERENTIAL LOCATION

The final-drive and differential (normally attached to the larger of the two gears of the final drive) may be located in different positions on the vehicle, according to 'chassis' design.

Complete the sketches below to show FOUR different types and state a typical example of each layout.

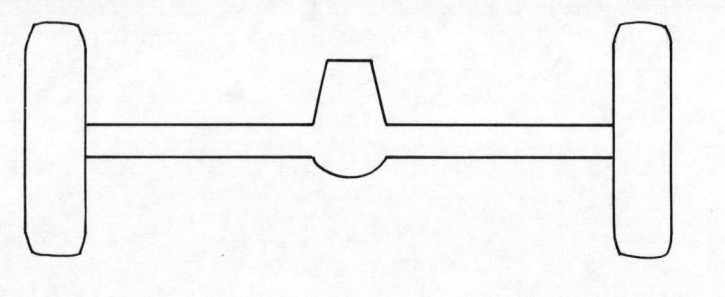

REAR MOUNTED LIVE AXLE

Make .... Model ....

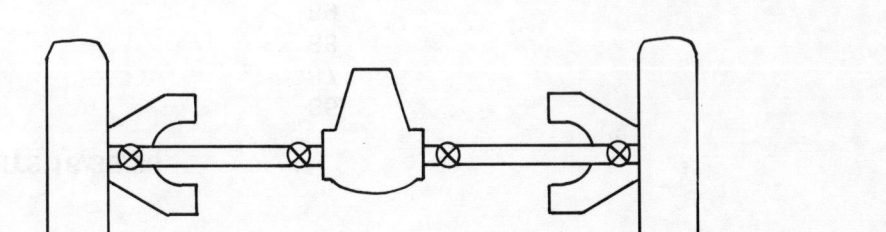

REAR MOUNTED TRANSAXLE

Make .... Model ....

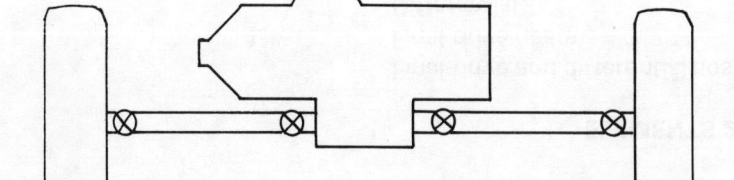

FRONT WHEEL DRIVE

Make .... Model ....

REAR DRIVE WITH IRS

Make .... Model ....

86

FINAL-DRIVE GEARS

The pair of gears providing the final drive are usually a bevel pinion and crown wheel or a worm and wheel. In many front-wheel-drive arrangements the final drive consists of a pair of helical gears similar to those used in most gearboxes. State the purpose of the final drive gears:

....
....
....
....

The drawings at A, B and C opposite show three types of final-drive gears used on motor vehicles.

Name the types shown and state alongside each drawing the relative advantages of each type of gear.

State the causes of axial thrust on the pinion and add arrows to the drawings to show the direction of axial thrust on drive and overrun for each type of gear:

....
....
....
....
....

Examine spiral bevel, hypoid and worm-and-wheel final-drive assemblies, count the gear-tooth numbers for each pair and state the gear ratios

Spiral bevel: ....

Hypoid ....

Worm and wheel: ....

Type: ....
....

A

Type: ....

B

Type: ....

C

D   Make a sketch at D to show the final drive gears for a front-wheel-drive transverse engine car and give a typical gear ratio for this arrangement.

Ratio: ....

DIFFERENTIAL

The purpose of the differential is to transmit equal driving torques to the half shafts while allowing the shafts to rotate at different speeds when the vehicle is travelling in other than a straight line.

The drawing below shows a final drive and differential assembly housed in an axle casing.

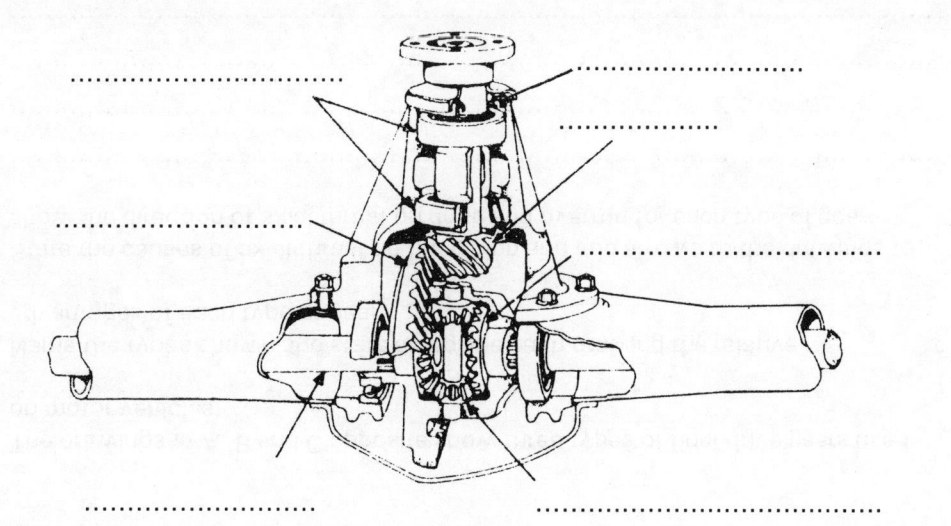

This type of axle carries the half shafts, hubs and brake assemblies and contains the lubricant. What other purpose on the vehicle does the type of axle fulfil?

....
....
....
....
....

OPERATION OF THE DIFFERENTIAL

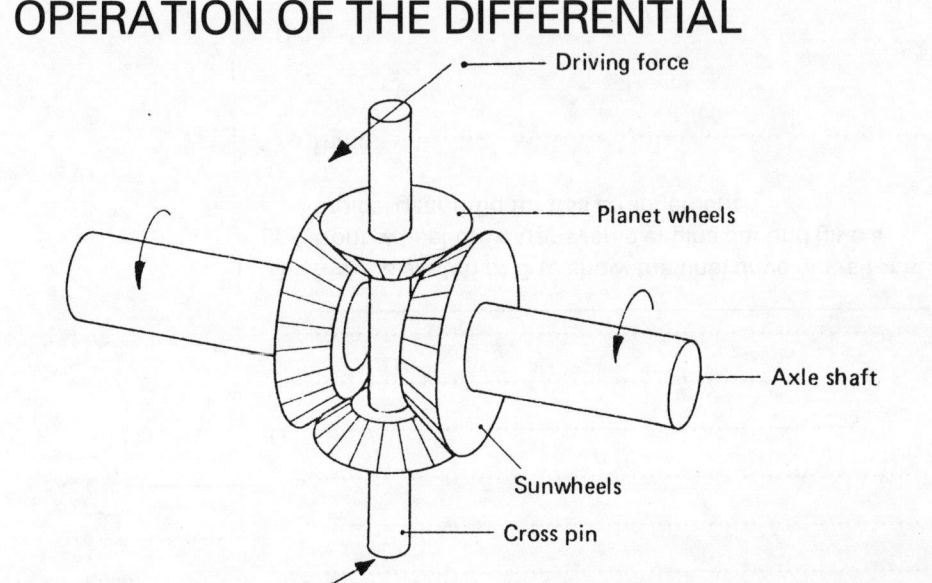

It can be seen from the drawing above that the planet wheels are pulled round by the cross pin. If the resistance to rotation at each axle shaft is equal, the sunwheels will rotate at the same speed as the cross pin, that is there is no relative rotation between planet wheels and sunwheels. Under what conditions would this occur?

....

INVESTIGATION

Observe the action of a differential by rotating the sunwheels at different speeds to each other and describe the action of the planet wheels.

....
....
....

By the action of the planet wheels rotating on the cross pin the sunwheels must rotate at different speeds (cornering), while still receiving equal driving torque from the planet wheels.

23.4 FINAL-DRIVE AND DIFFERENTIAL LOCATION

The entire final-drive and differential assembly runs in four bearings housed in the axle casing (two on the pinion shaft and one on either side of the crown wheel and differential assembly).

By examining a rear-axle assembly, show on the drawing below the position and type of bearings used to support the final drive.

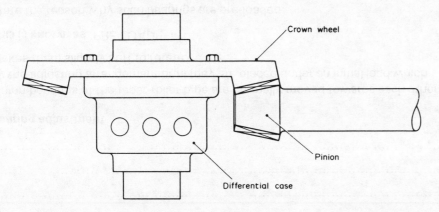

- Crown wheel
- Pinion
- Differential case

Add centre lines to the drawing below to illustrate the geometric pinciple upon which the taper roller bearing operates.

Taper roller bearing

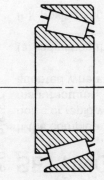

Straddle mounting of the pinion is common practice in hgv applications; make a simple sketch to illustrate this arrangement and state briefly the reason for its use.

....

....

....

....

When a large-diameter crown wheel is employed (hgv) a thrust block is often used to prevent any sideways deflection. Add a crown wheel thrust block to the drawing below; include the method of adjustment.

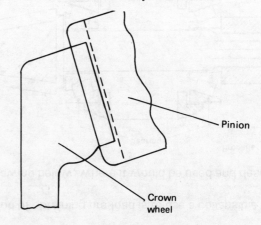

- Pinion
- Crown wheel

89

SETTING-UP AND ADJUSTMENT

If the complete final drive assembly is to operate satisfactorily, without undue noise or rapid wear, the clearances and adjustments must be as specified by the manufacturer. To meet these specifications a definite procedure must be adopted when:

(a) adjusting the bearings

(b) positioning the final-drive gears to obtain correct meshing.

Briefly outline such a procedure.

....
....
....
....
....

Bearing adjustment

Pinion bearings of the taper-roller type are normally moved towards each other by the adjusting arrangement until they are placed under an initial load which makes them slightly stiff to rotate.

This is known as 'PRE-LOAD'.

State the reason why such bearings are pre-loaded.

....
....
....

Name the types of bearings that should not be pre-loaded.

....
....
....

A popular method of obtaining pre-load is to use a collapsible spacer.

Show on the drawing below, where it would be used and describe how it would be used.

....
....
....

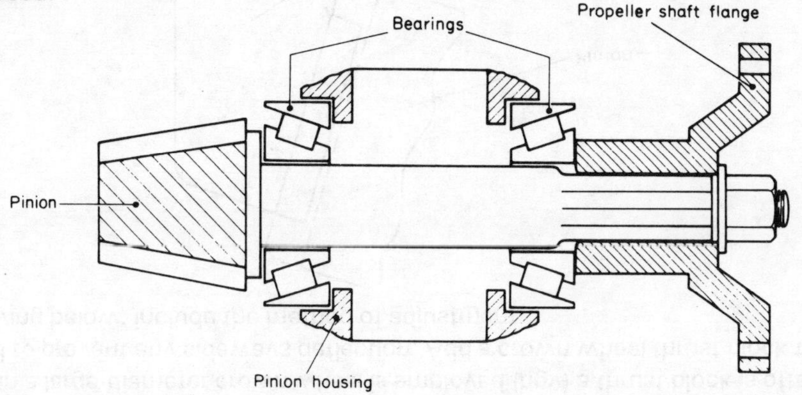

The sketch below shows a popular type of pre-loading gauge. Describe how it would be used.

....
....
....

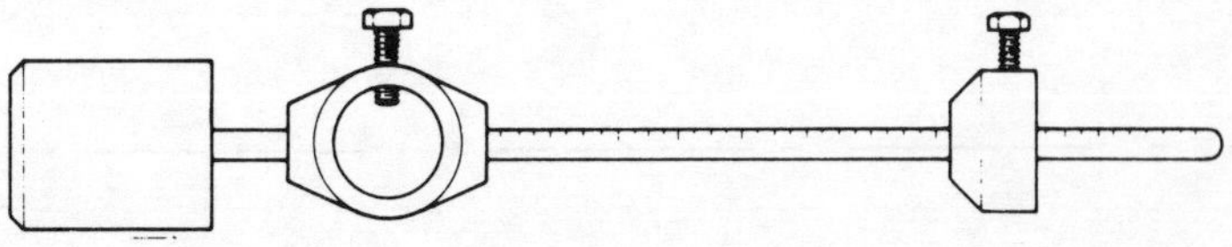

Tooth meshing

The correct meshing of the final-drive gears is obtained by moving the crown wheel or pinion in or out of mesh until the correct position is found.

Complete the sketch to show how the correct backlash for the final-drive gears can be checked by using a clock gauge (or dti) assuming that the pinion is held steady.

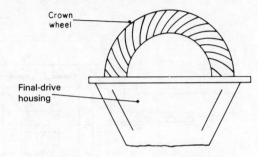

Crown wheel

Final-drive housing

Final tooth meshing can be checked by smearing the gear teeth with marking compound. Sketch the tooth marking, for the meshings indicated, on the teeth below.

CORRECT MESHING   TOO FAR IN MESH   TOO FAR OUT OF MESH

State the reason why final-drive gears are supplied in mated pairs.

....

....

....

....

....

....

INVESTIGATION

1. Remove the final-drive assembly (either bevel gear or worm and wheel) from an axle casing.

   Describe the method of pre-loading the bearings.

2. Pinion bearings:

....

   Side bearings:

....

3. State the backlash setting and the pre-load setting for the assembly.

   Backlash ....   Pre-load (pinion ....)

   (side bearings ....)

4. Assemble the unit and describe briefly the procedure for adjusting the final-drive gears to the manufacturer's specifications.

   A. ....

   ....

   ....

   B. ....

   C. ....

   ....

   D. ....

   ....

   E. ....

LIMITED-SLIP DIFFERENTIAL

Sports and racing cars are vehicles with high power-to-weight ratios and as such can, even on good surfaces, cause a driving wheel to spin, say, when cornering, owing to the high power available and weight transfer.

The limited-slip device is a device incorporated in the differential, which automatically applies a brake to the spinning half shaft, thereby maintaining a torque in the other half shaft.

The limited-slip device shown below incorporates spring-loaded cone clutches behind each sunwheel. Should one half-shaft/sunwheel spin, the torque required to slip the cone clutch will be transferred to the other non-slipping half shaft.

Label the drawing and add arrows to indicate power flow.

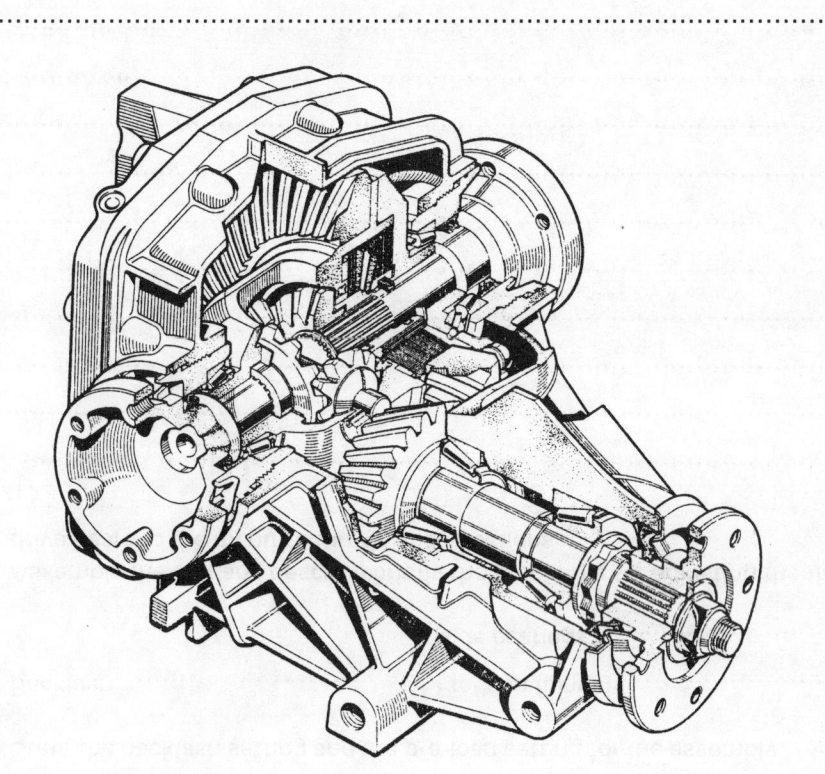

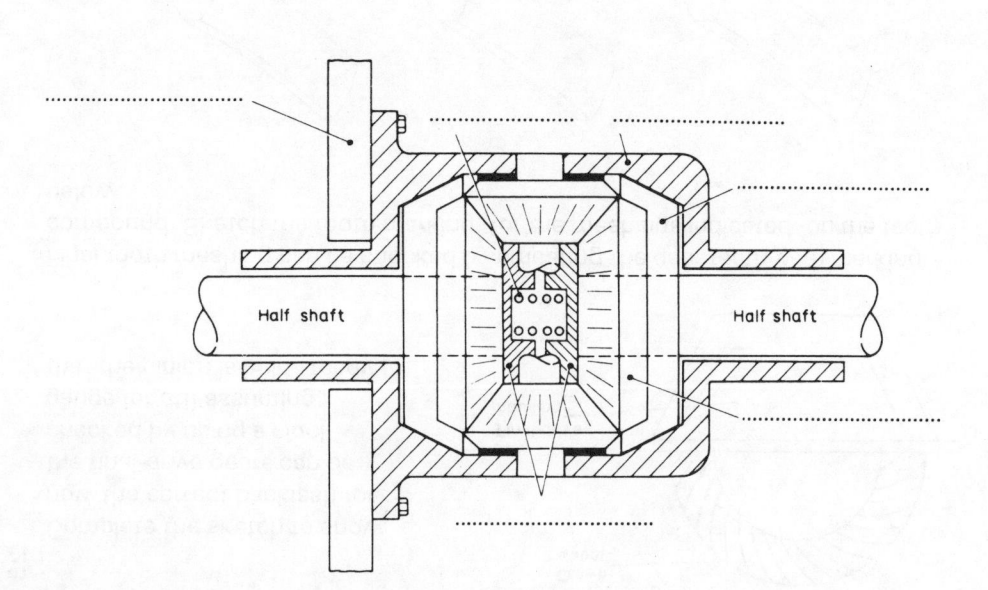

Half shaft   Half shaft

The type shown above is one of many different designs of limited-slip differential. The most popular type in use today is shown top right. Name the type shown and label the drawing.

Describe briefly how to check a limited-slip differential for operation and state any safety precautions to be observed when working on a vehicle fitted with a limited-slip differential.

....

....

....

....

....

....

TANDEM (TWIN) DRIVE AXLES

Many hgvs have two 'closely spaced' 'live' axles providing the drive, each axle having a differential to allow for wheel speed variations on that axle.

The drawing at (a) below shows a worm and wheel arrangement. Complete the drawing at (b) to show a tandem drive which uses crown wheels and pinions.

(a) Worm and Wheel Tandem Drive

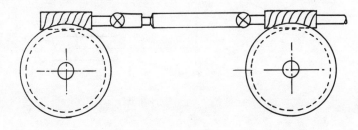

(b) Bevel Gear Tandem Drive

When a tandem drive is employed the forward axle differential will allow for speed variations of its inner and outer wheels and the rear-axle differential will do likewise. It is, however, desirable to allow for a speed variation between the two final-drive crown wheels or worm wheels. Why is this?

....

INTER AXLE (THIRD) DIFFERENTIAL

If the drive to the two final-drive pinions or worm gears is transmitted through a third differential, speed variation between the two sets of final-drive gears can occur and the driving torque is shared equally between the two axles.

The simplified drawing below shows a worm and wheel tandem drive incorporating a THIRD differential. Describe the operation of this system.

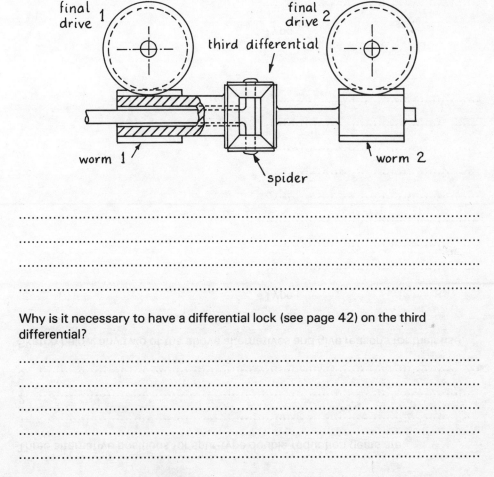

....
....
....
....

Why is it necessary to have a differential lock (see page 42) on the third differential?

....
....
....
....

DOUBLE-REDUCTION GEARS

A double-reduction gear used in conjunction with the final-drive gears provides, in two 'stages or steps', a large gear reduction at the driving axle.

State why double-reduction gears are used?

....

....

....

....

....

....

Complete the drawing below to show a spur-gear double-reduction arrangement.

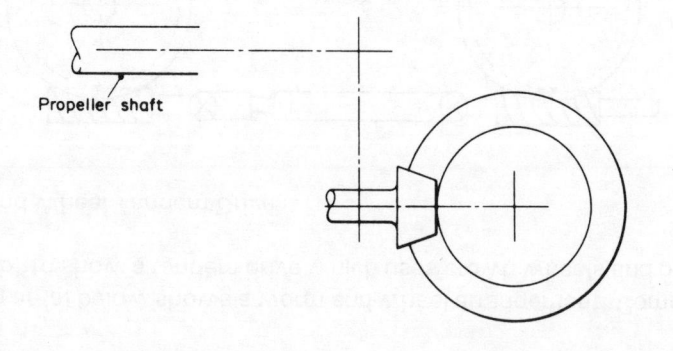

Propeller shaft

Three alternative positions for spur-type double-reduction gears are:

1. ....

2. ....

3. ....

Sketch below any two of the above alternatives and give reasons for their use.

Type

....

....

....

....

....

....

....

....

Type

....

....

....

....

....

....

....

Epicyclic double-reduction gears

The drawing below illustrates an epicyclic reduction gear in the final-drive assembly; describe its operation.

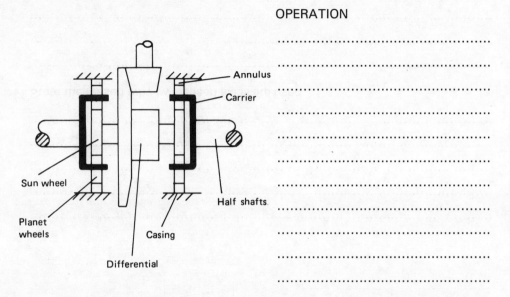

Annulus

Carrier

Sun wheel

Planet wheels

Casing

Differential

Half shafts

OPERATION

....
....
....
....
....
....
....
....
....
....

Examine an epicyclic hub-reduction gear and complete the drawing below to show the arrangement.

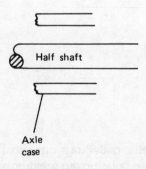

Half shaft

Axle case

State the main advantage of the epicyclic-type reduction gear.

....
....
....
....
....

TWO-SPEED AXLE (EPICYCLIC)

The drawing below is a section through a two-speed axle employing an epicyclic gear. Describe the operation of the axle as it is shown, that is, in low ratio.

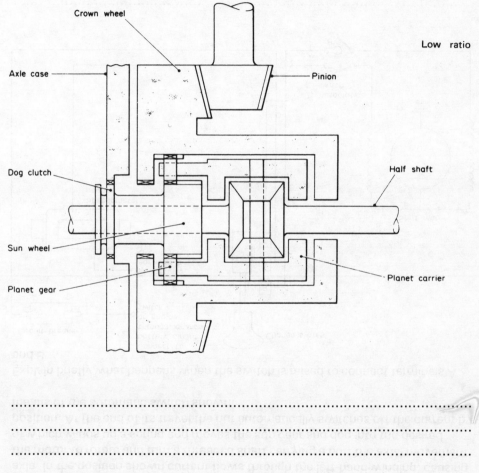

Crown wheel

Axle case

Low ratio

Pinion

Dog clutch

Sun wheel

Planet gear

Half shaft

Planet carrier

....
....
....
....
....
....

Complete the drawing below to show the position of the sliding dog clutch when high ratio is engaged, and describe the operation in high ratio.

Axle case

Crown wheel

....

....

....

....

....

State the reason why two-speed axles are used.

....

....

....

....

An electrically operated shift system is shown opposite; name one other shift system.

....

TWO-SPEED AXLE SHIFT UNIT

The diagram shows an electrically operated shift unit for an Eaton two-speed axle. In the position shown current flows through the left-hand winding, causing the motor to rotate and turn a threaded shaft carrying a nut, the axial movement of which winds up a spring and moves the sun gear and dog into the desired position. At the end of its travel the nut automatically switches off the current by means of the automatic switch shown.

Explain briefly what happens when the switch is raised to connect terminals A and B.

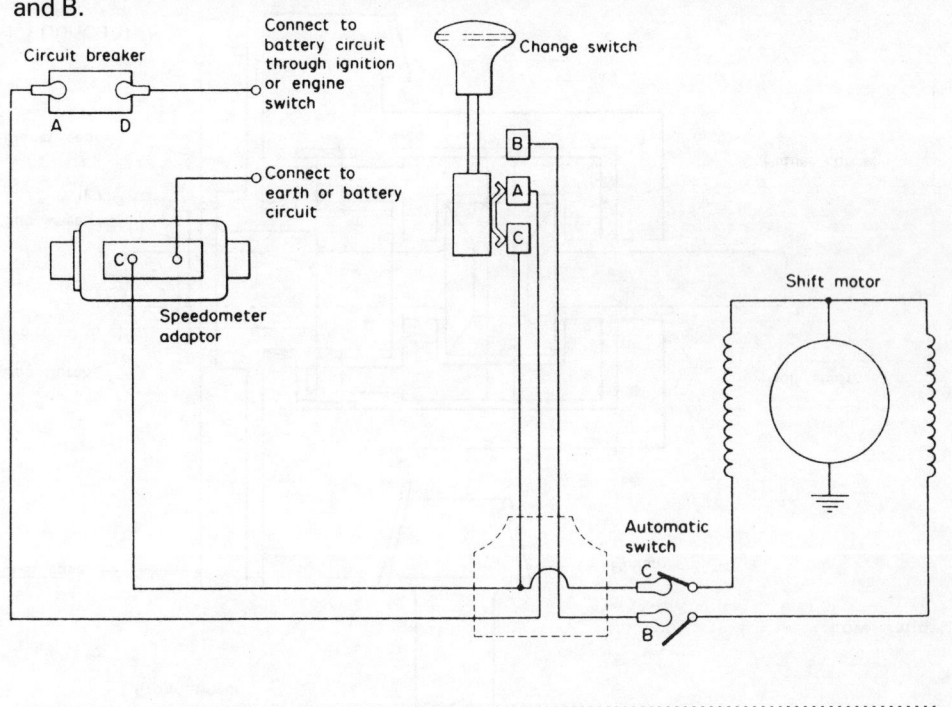

....

....

....

....

MAINTENANCE AND REPAIR

Briefly describe methods of protecting the final drive and differential system against the usual hazards during use or repair.

HAZARD	PREVENTIVE TREATMENT OR ACTION
Corrosion	
Mechanical damage during repair	
Lubricant contamination	
Pressure build-up	

The purpose of routine maintenance is to prolong service life and ensure reliability, safety and efficiency. List below FIVE maintenance items.

....

....

....

....

....

State general rules for efficiency and special precautions needed during maintenance and adjustments in respect of the items listed below.

ITEM	RULES/PRECAUTIONS
Vehicle lifting supporting and checking	
Support of engine gearbox and suspension	
Cleanliness: (a) general (b) lubricant spillage (c) friction faces	
Lubricants	
Brakes pipes and cables	

FINAL-DRIVE AND DIFFERENTIAL FAULTS, SYMPTOMS, PROBABLE CAUSES AND CORRECTIVE ACTION

Complete the table in respect of the faults listed.

FAULTS	SYMPTOMS	PROBABLE CAUSES	CORRECTIVE ACTION
Oil leakage			
Bearing failure			
Excessive end float in pinion bearings			
Damaged or broken final-drive gears			
Worn differential gears			
Faulty VC or worn clutch pack			
Spline damage			
Incorrect meshing of final-drive gears			

Chapter 6

Suspension

ELEMENTS 25/46/47   **UNITS 18/39/40**

SUSPENSION

The purpose of the suspension system on a vehicle is to:

1. Minimise the effect of road surface irregularities on passengers, vehicle and load.

2. ....

3. ....

4. ....

5. ....

6. ....

Many different forms of suspension are employed on road vehicles — some very simple and relatively inexpensive, others highly sophisticated and expensive. Suspension systems fall into one of two main categories: Independent and Non-independent systems.

Describe with the aid of simple diagrams the essential difference between the two categories.

....

....

....

....

....

....

Leaf spring suspension

The leaf spring is widely used on light and heavy goods vehicles and is still in limited use on car rear suspension. Used in conjunction with a beam axle the leaf spring is a simple NON-INDEPENDENT suspension system. The length, width, thickness and number of leaves varies according to the load requirement. An advantage of the leaf spring is that it can provide total axle location in addition to its springing properties.

Complete the drawing below to show a multi-leaf suspension spring.

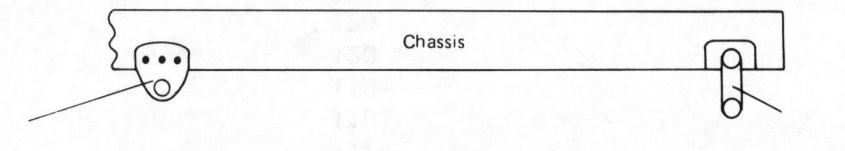

The leaf spring is positively located on the axle by a ....
and secured to the axle by .... As the spring flexes
the distance between the spring eyes, that is, the length of the spring, varies.
This is accommodated by the .... or, in some cases,
an open-ended spring operates in a ....
....

Overloading of the main leaf on rebound is prevented by ....
....

In relation to suspension, state the meaning of:

SPRUNG WEIGHT ....

....

UNSPRUNG WEIGHT ....

....

SUSPENSION DAMPERS

Contrary to popular belief, the function of the suspension damper is not to increase the resistance to spring deflection but to control the oscillation of the spring. The basic principle of hydraulic dampers is that of converting the energy of the deflecting spring into heat. This is achieved quite simply by pumping oil through a small hole.

The graph below shows the oscillations of an 'undamped' spring as it dissipates energy following an initial deflection.

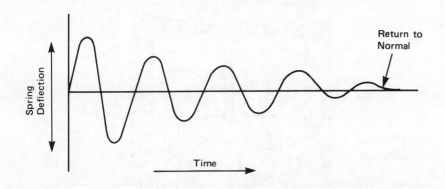

Draw a graph below to illustrate the effect of a damper on the oscillation of a spring.

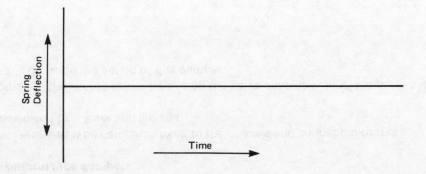

Direct-acting hydraulic damper

The telescopic type, as this type is known, is located directly between the frame and suspension unit or axle, thus eliminating the need for levers or links. The principle of operation is the transfer of fluid through small holes from one side of the piston to the other. Complete the sketch below by adding a sectioned view of the interior of the damper. Label the essential parts. Alongside it, sketch in enlarged detail the piston and valve assembly.

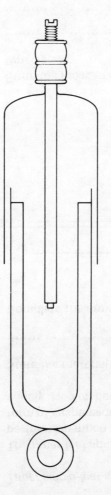

State the purpose of the fluid reservoir:

....
....
....
....
....
....

101

Lever-arm type damper

The lever-arm type damper is fixed to the chassis and connected to the suspension by a lever and linkage.

Complete the sketch below by adding the valve assembly, label the parts and briefly describe the action of this damper.

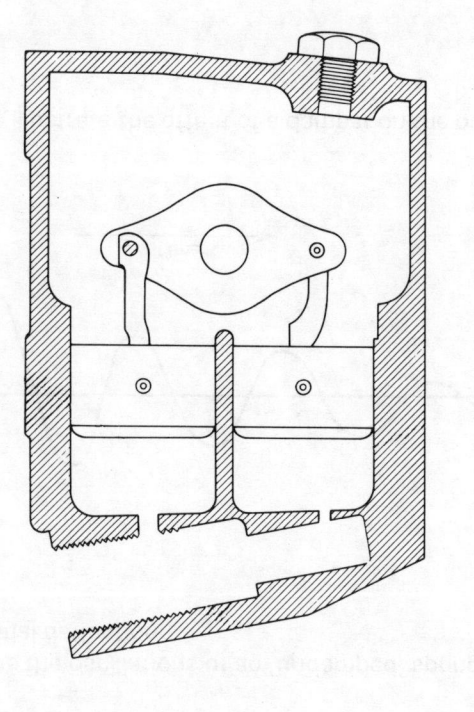

....

....

Taper single-leaf spring suspension

The taper-leaf spring that increases in thickness towards the centre and performs in much the same way as a multi-leaf spring; it is an extremely popular form of suspension on hgv trailers, where it is in open-ended form. It is also widely used on beam-type car and light commercial rear axles.

State the material from which the spring may be made:

....

Complete the drawing below to show a taper-leaf spring suspension.

Chassis

State the main advantage of the taper-leaf spring compared to the multi-leaf-type spring:

....

....

....

Why are multi-leaf and single leaf springs made thicker towards the centre?

....

....

COIL SPRING SUSPENSION

The coil spring is used on many car front and rear suspension systems. As with the single-leaf spring this type has no self-damping effect. The rate of oscillation can be controlled by the thickness of the spring and its length.

Show, on the three independent front suspension arrangements below, the position of the coil spring.

TORSION BAR SUSPENSION

The torsion bar is, in effect, a coil spring which has been straightened out.

Label the drawing and describe the operation of the torsion bar suspension shown.

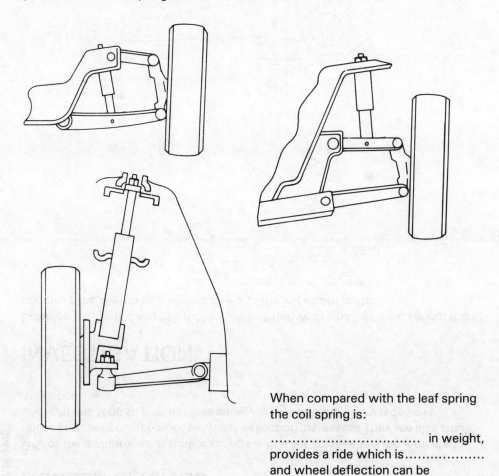

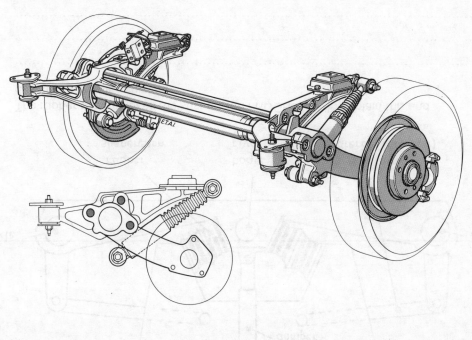

When compared with the leaf spring the coil spring is:

.... in weight,

provides a ride which is..................

and wheel deflection can be

....

Operation

....

....

....

....

RUBBER SPRINGS

One of the advantages of rubber springs is that for small wheel movements the ride is soft, becoming harder as wheel deflection increases. They are also small and light and tend to give out less energy in rebound than they receive in deflection.

INVESTIGATION

Examine the front suspension of a BL Mini fitted with a rubber cone spring, then complete the sketch below. Add labels to the important parts.

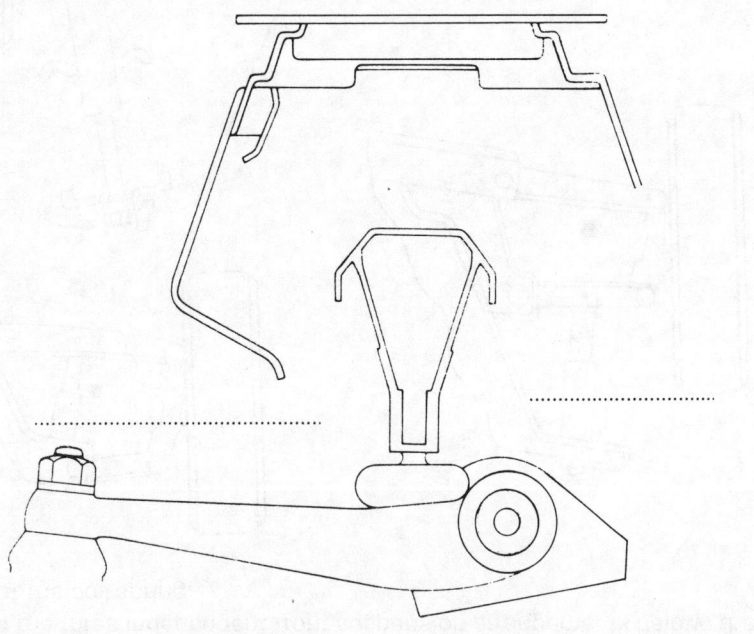

....

....

The deflection of the rubber spring in the suspension shown above is small when compared with the actual rise and fall of the wheel. State how this is achieved.

....

....

....

Heavy goods vehicle application

The trailer suspension shown below employs rubber springs. The particular benefits offered by rubber springs in this application are: weight-saving and ride improvement.

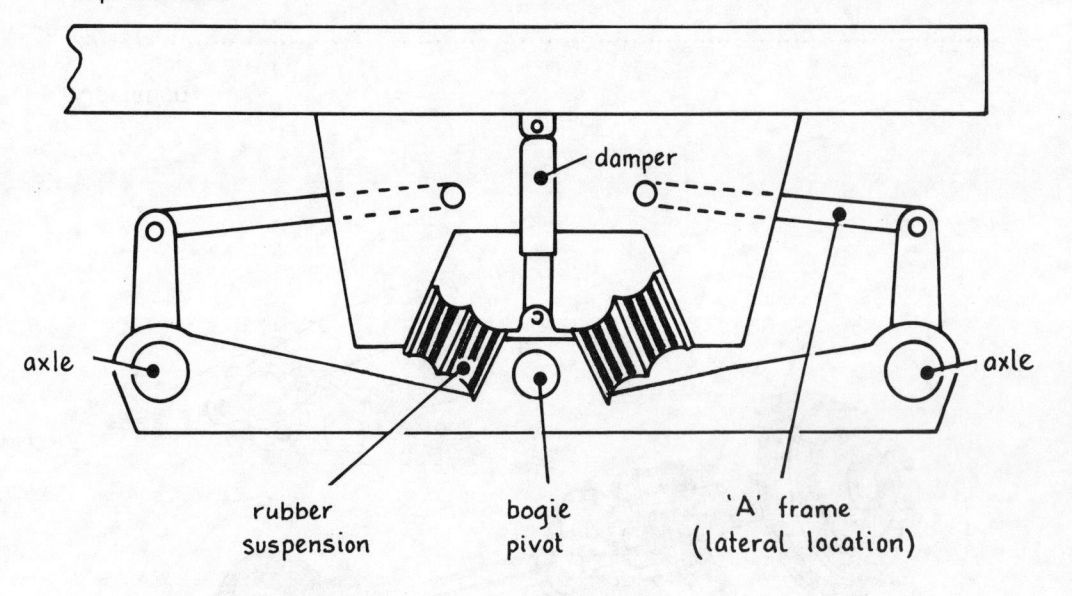

The rubber spring makes the suspension PROGRESSIVE. Explain this and describe how it is achieved.

....

....

....

....

....

....

....

HYDROLASTIC SUSPENSION

This combination of rubber springing and hydraulic pressure was developed by the British Motor Corporation (now the Austin Rover Group Ltd) and is a development of the Mini-type rubber suspension. To each suspension arm is fitted a hydrolastic unit which consists of a steel cylinder mounted to the underbody of the car. A tapered piston, complete with a rubber and nylon diaphragm connected to the suspension arm, fits at the lower end of the cylinder. When the wheel is deflected the piston moves upwards and via fluid action compresses the rubber spring, at the same time displacing fluid into the hydrolastic unit at the opposite end of the vehicle.

The drawing below shows the arrangement for one side of the vehicle; label the drawing and describe, under the headings at the top opposite, the action of the system during 'pitch' and 'roll'.

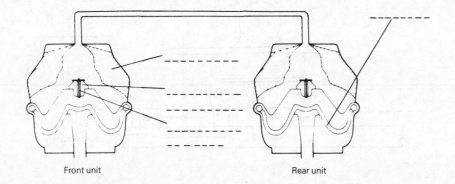

Front unit   Rear unit

State the composition of the fluid used in this type of system:

....

The interlinked Hydrolastic and Hydragas suspensions are no longer in use on current production models, although it is a syllabus requirement. However the Hydragas displacer units are used independently in conjunction with separate damper units.

Pitch

....

....

....

....

Roll

....

....

....

....

HYDRAGAS SUSPENSION

The hydragas system is a development of the hydrolastic system. The main difference is that the rubber spring is replaced by a pneumatic spring.

Complete the suspension drawing below by adding the hydragas displacer unit.

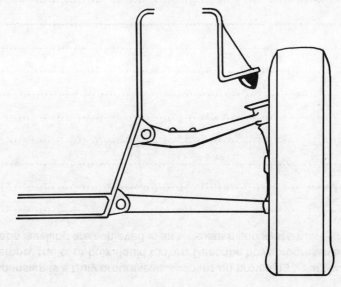

AIR SUSPENSION

Air springs usually consist of reinforced rubber bellows containing air under pressure, situated between the axle, and an air capacity tank mounted to the chassis. They have an important advantage in that by fitting levelling valves to each wheel unit the height of the vehicle can be kept constant whatever the load.

The main features of air suspension are shown below; label the drawing.

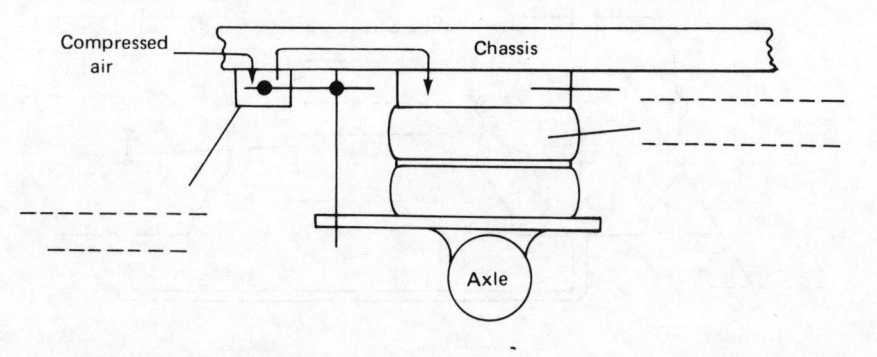

Air suspension is very popular on hgv trailers, particularly tankers. Complete the drawing below to show the layout of an air suspension system for a tandem-axle trailer; label all parts.

Chassis

Air suspension is a truly progressive suspension providing a soft-cushioned ride for an empty trailer or bulk-liquid tanker. Describe how progressiveness and automatic levelling are achieved in an air suspension system.

....
....
....
....
....
....
....
....
....
....
....
....
....
....

Advantages of air suspension are:

less trailer hop when empty
less body/chassis maintenance required
constant platform height (advantageous when using stacker trucks from a
     loading bay)
less weight

One drawback is the increased body sway on high, loaded vehicles.

HYDRO-PNEUMATIC SUSPENSION

As with air suspension, the spring is pneumatic. The main difference is that a fixed quantity of gas (nitrogen) is contained in a variable-sized chamber; liquid is used to transmit the force from the suspension link to the gas (via an intervening diaphragm).

Complete the drawing below to illustrate the principle of operation of a hydro-pneumatic suspension system.

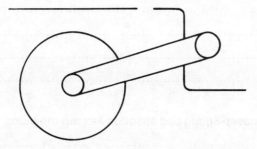

One system of hydro-pneumatic suspension incorporates a levelling valve and a hydraulic accumulator. In this system the vehicle maintains a constant ground clearance irrespective of load.

The driver can also adjust the vehicle height or ground clearance to any one of three positions.

Give examples of vehicles that use this form of suspension.

....

....

....

....

The simple layout below shows a suspension system with a levelling valve and an accumulator.

Add arrows to the hydraulic circuit to indicate the direction of fluid flow, label the drawing and describe how the system operates.

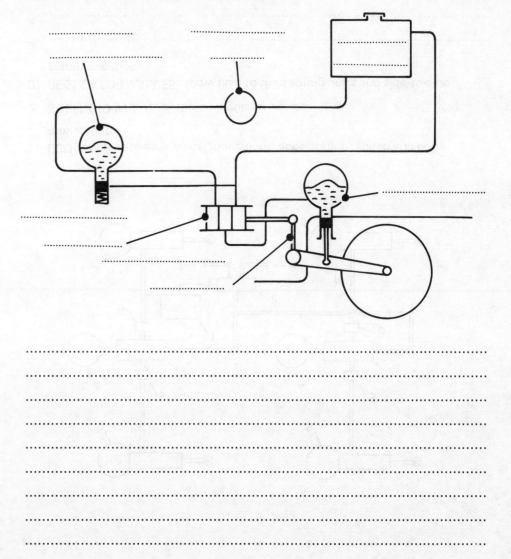

....

....

....

....

....

....

....

....

'ACTIVE' SUSPENSION

A development of the hydro-pneumatic suspension is shown in the diagram opposite.
Additional 'THIRD' spring units are added between the front and rear pairs. This system provides a variable spring rate and roll stiffness, that is, the suspension is ACTIVE.

What are the benefits of ACTIVE suspension?

....

....

....

....

....

....

Study the diagram, complete the key opposite and briefly describe the action of this system.

....

....

....

....

....

....

....

....

....

....

....

....

1. ECU: senses steering wheel movement, acceleration, speed and body movement.

2. SOLENOID VALVE: actuates regulator valves.

3. REGULATOR VALVES: allow fluid to third spring units and side-to-side communication.

4. ....

5. ....

6. ....

7. ....

8. ....

9. ....

UPRATED SUSPENSION

If the suspension springing on a vehicle is uprated it offers a greater resistance to deflection, it is stiffer and will more easily cope with heavier loads.

Give reasons for using uprated springs:

1. ....

2. ....

One method of uprating or reinforcing the suspension springs is to replace the dampers with spring damper units (a) opposite, in which a variable rate coil spring increases its resistance according to the additional weight carried. The damper itself would be stiffer than the standard unit. Make a sketch at (b) to show an alternative method of uprating the springs.

Gas pressurised dampers

The simplified drawing opposite shows a gas damper in which a floating piston separates the gas from the fluid. This is a single or mono-tube design (no fluid reservoir).

State two functions of the gas:

1. ....

....

2. ....

....

....

....

Note: In some gas dampers the gas is not separated from the fluid and an emulsion is formed which performs in much the same way.

State the advantages of the mono-tube design.

....

....

....

(a) 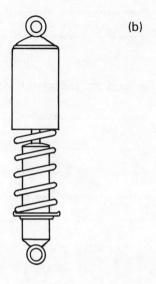 (b)

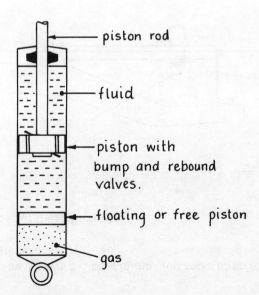

piston rod

fluid

piston with bump and rebound valves.

floating or free piston

gas

Ride leveler system

For vehicles towing trailers, caravans etc. or where a good ride is essential with a variety of loads (such as an estate car), it is desirable to uprate the rear springs.

In addition to this, certain manufacturers offer height regulating 'shock absorbers' (dampers) which will raise the loaded rear end of a vehicle back to its unloaded height.

Make a simple diagram to illustrate such a system and describe its operation.

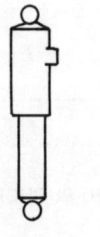
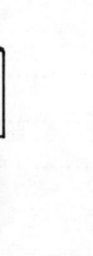

....

....

....

....

....

What benefits does this system offer?

....

....

....

....

Bump and rebound stops

The drawing below shows a rubber bump stop and a strap to limit rebound in a suspension system.

State the purpose of these fitments.

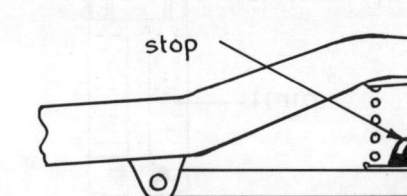

Purpose

....

....

....

....

Trim height

What is meant by 'trim height' (or standing height) and how is it measured?

....

....

....

....

UPRATED DAMPERS

How is the operation of a damper affected if it is 'UPRATED'?

....

Give TWO reasons for the use of uprated dampers:

1. ....

....

2. ....

....

On some dampers the rate can be varied by making a manual adjustment on the damper itself. With some installations a solenoid-operated adjustment can be activated by a switch on the dashboard. This type of damper is popular on vehicles which are 'driven hard' on occasions. (Rally type work). The damper rate can be adjusted to suit the usage of the vehicle or adjusted to satisfy the handling/ride characteristics desired by the driver.

With the aid of a sketch describe a damper incorporating a manual rate adjuster.

Adjustment:

....

....

....

....

VARIABLE RATE DAMPERS

The simplified damper shown is automatically uprated as the load increases. Add arrows to the drawing at (a) to indicate fluid flow during light load and complete the sketch at (b) to show high load operation.

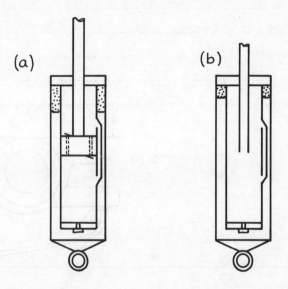

(a)   (b)

Under what other operating conditions would this type of damper offer increased damping?

....

....

....

Why does the damper rate or its resistance to movement increase under load or at extreme bump or rebound?

....

....

ACTIVE SUSPENSION CONTROL

In an ideal suspension system, the spring and damper rate are automatically adjusted during vehicle operation to ensure that the ride and handling standards are optimum under all operating conditions.

In the system below, a microcomputer controls the spring rate, damper rate and ride height of the rear suspension. This system can be adapted to provide control for both front and rear suspension.

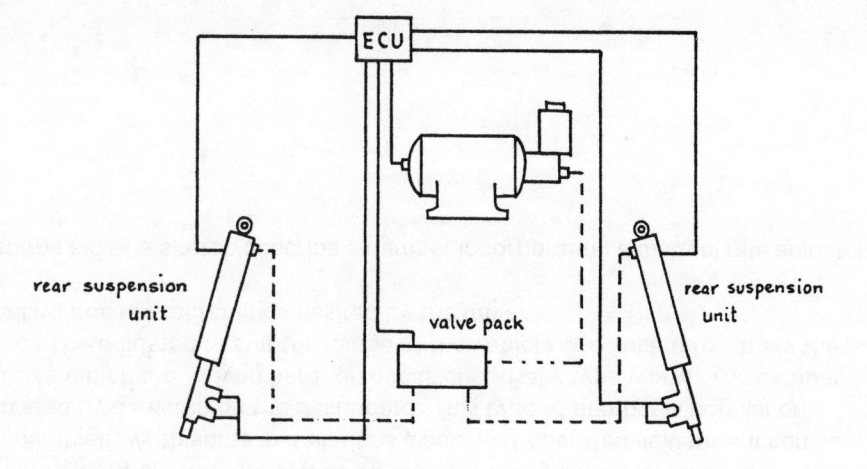

Describe the operation of the system:

....
....
....
....
....
....
....
....

CHECKING DAMPER OPERATION

Bounce test

This involves pushing down on the car body at each corner and noting the oscillations of the vehicle before it again becomes stationary in its normal level position. The 'rule of thumb' guidance (MUNROE) is that a vehicle taking more than one and a half oscillations has ineffective damping of the spring.

Other methods involve the use of equipment which will produce a graphical record of vehicle oscillation.

Examine one type of damper test equipment (Drop test or Eccentric roller) and describe the system.

....
....
....
....
....
....
....
....

AXLE LOCATION

Coil, torsion bar, rubber and air springs serve in general merely to support and control the vertical load and do not locate the axle or wheel assembly in any way.

Additional arms or links must be provided to locate the wheels both fore and aft and laterally, and to resist the forces due to drive and brake torque reaction.

Axle location (ifs)

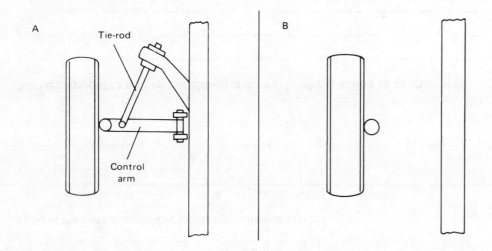

The tie-bar shown at A, above, locates the wheel assembly to prevent rearward movement of the wheel when the brakes are applied, when the vehicle accelerates or when the wheel strikes a bump.

Sketch an alternative method of providing this location B.
On some suspension systems the anti-roll bar serves also as a tie-bar.

State the meaning of the term 'compliance' in relation to suspension.

....

....

Many cars are fitted with a rear beam axle and coil springs. Complete the drawings below to show axle location by RADIUS ARMS and PANHARD ROD.

Radius arms

Radius arms provide fore and aft location and prevent axle casing rotation due to drive and brake torque reaction.

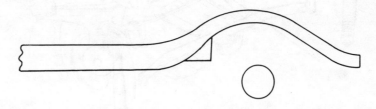

Panhard rod

A Panhard rod gives lateral, that is, sideways, axle location.

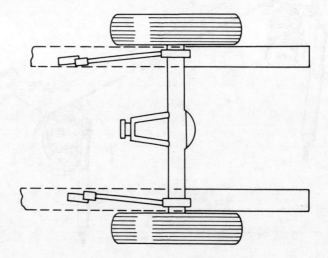

Put appropriate descriptive headings above the suspension illustrations at (a) and (b) opposite and label the drawings.

The TORSION BEAM, as shown at (a), is now widely used. What is its purpose?

....
....
....
....
....

How is the wheel assembly located in suspension (b)?

....
....
....
....
....

Name component arrowed on drawing (b) and briefly describe its function and action.

....
....
....
....
....

The suspension/subframe assembly is attached to the chassis/body through bonded rubber mountings; indicate on the drawings where such mountings are positioned.

Why is a suspension subframe considered desirable?

....
....
....

(a)

(b)

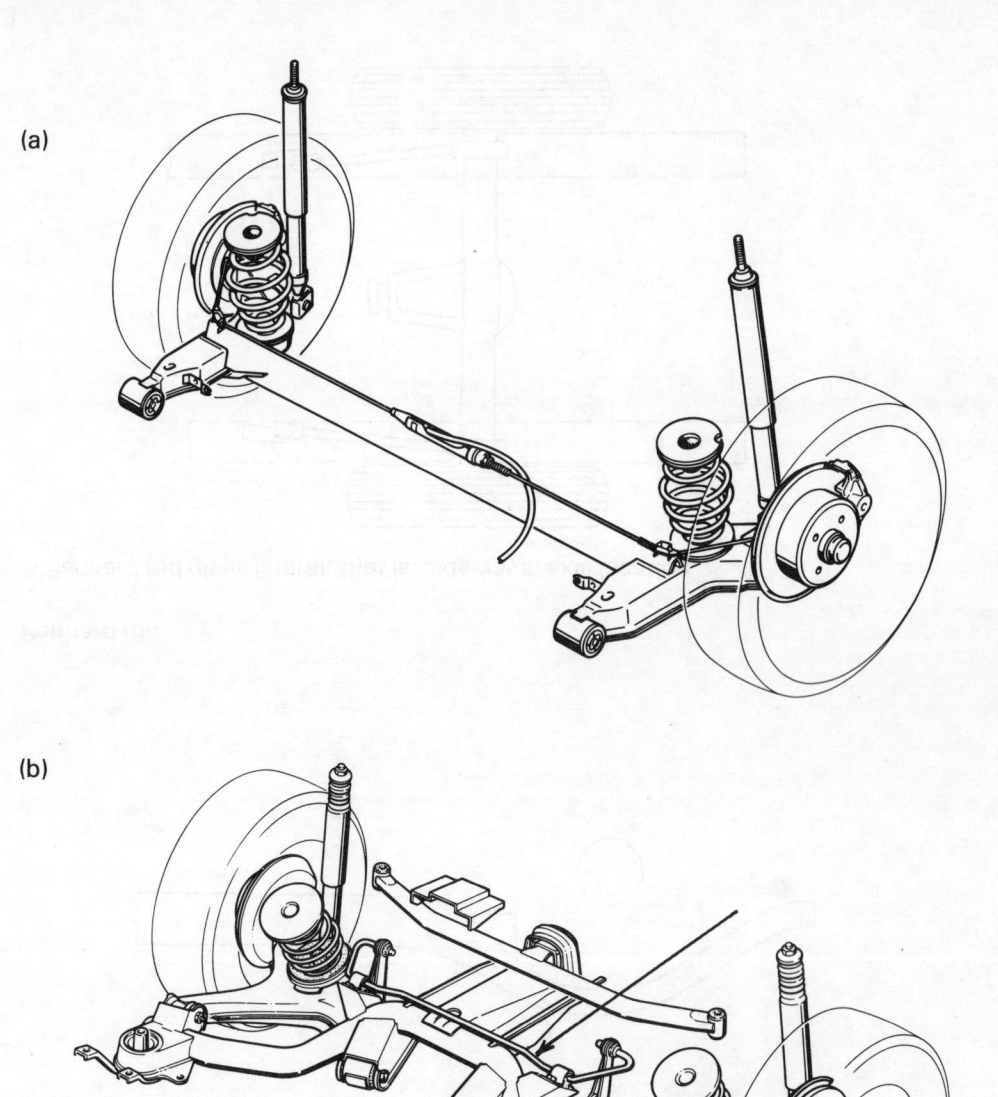

114

DE-DION AXLE

The De-Dion axle is a compromise between the live beam axle and independent suspension. The final drive and differential are mounted on the chassis and the road wheel assemblies are carried on a beam type dead axle. Coil springs are used and the axle is located by trailing links and a Panhard rod. In some systems a Watts linkage provides lateral location rather than a Panhard rod.

Complete the drawing below to show the De-Dion system.

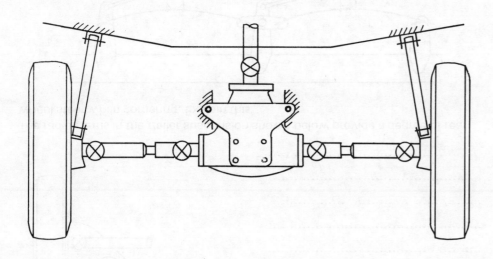

State two advantages of this layout;

1. ....

2. ....

Name two vehicles that use the De-Dion layout:

Make .... Model ....

Influences of load variations on wheel geometry and alignment

The wheel alignment and suspension geometry can vary as the wheels rise and fall to follow the road contour, and can be affected by load. The degree to which alignment and geometry does alter depends largely on the actual suspension design and the amount of suspension movement involved.

Describe how suspension geometry is affected when the suspensions named below are involved.

Parallel arm type

....
....
....
....
....

Unequal length wishbone type

....
....
....
....
....

Macpherson type

....
....
....
....

REAR WHEEL 'SUSPENSION STEER'

The semi-trailing link IRS shown below produces a steering effect when the vehicle is cornering. Describe this.

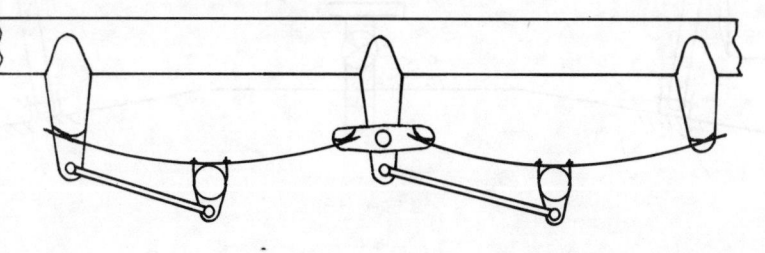

....

....

....

....

....

....

....

....

The radius arms in the trailer suspension shown below provide a degree of rear wheel steer when cornering. Explain this.

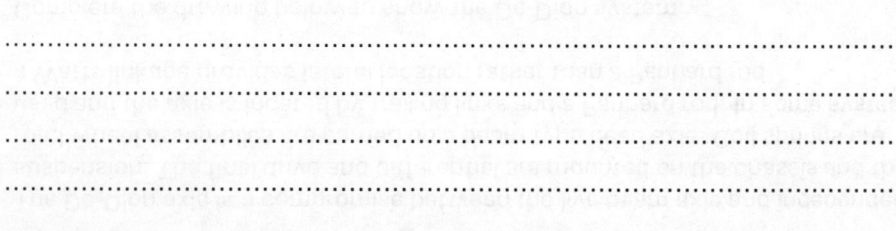

....

....

....

....

....

ENERGY CONVERSION

When a spring is deflected, for example

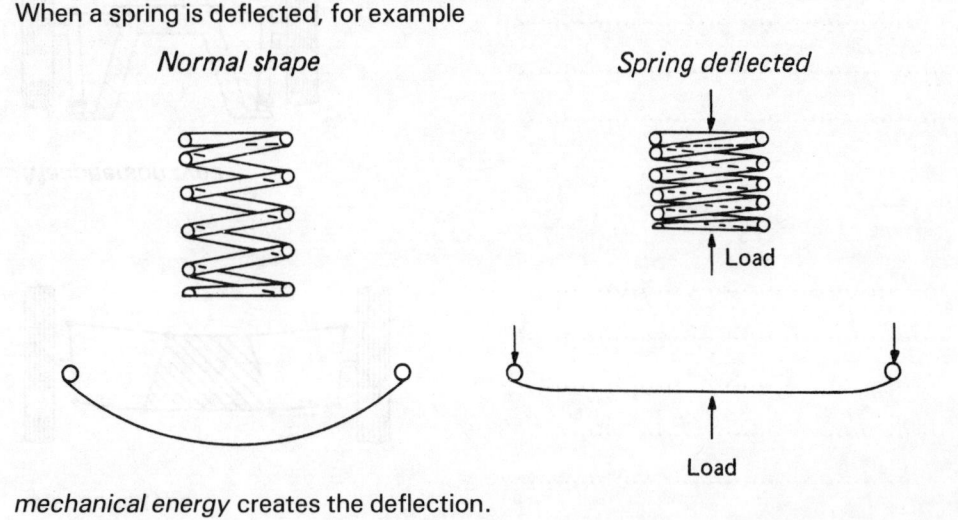

Normal shape   *Spring deflected*

Load

Load

mechanical energy creates the deflection.

(a) What happens to the energy in an 'undamped' spring?

....

....

....

....

....

(b) What happens to the energy put into the spring when a hydraulic damper is used to prevent excessive spring oscillation?

....

....

....

....

....

LEAF SPRING – COMMERCIAL VEHICLE

A modern hgv TAPER leaf spring suspension is shown below. Label the drawing and state the purpose and describe the action of the lower two leaves.

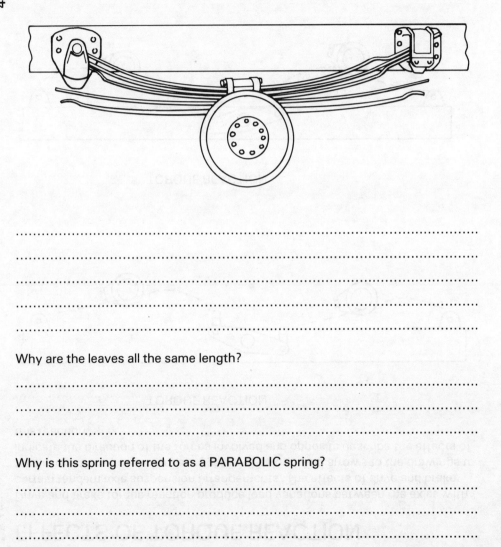

....

....

....

....

Why are the leaves all the same length?

....

....

....

Why is this spring referred to as a PARABOLIC spring?

....

....

....

MULTI-AXLE SUSPENSION

When two 'closely spaced' axles are employed at the rear of a hgv, the suspension design must ensure that:

(a) the payload, irrespective of position, is shared by the two axles;

(b) the forces due to road shocks are balanced out across the two axles.

To satisfy the requirements (a) and (b) above, tandem-axle suspensions are interconnected. One of the most common, and simplest, methods is to employ a 'balance beam'.

Complete the drawing below to show a balance-beam arrangement and describe briefly (use arrows on the drawing) how load balancing is achieved.

....

....

....

....

....

One disadvantage of the above system is that when a relatively short balance beam is used, wheel movement is rather limited; increasing the length of the balance beam allows a greater wheel movement at the expense of an increased axle spread, which causes more tyre scrub.

EFFECTS OF TORQUE REACTION

Drive and brake torque reaction produce load variations between the axles with certain tandem axle suspension arrangements. The effects of drive and brake torque reactions are shown (exaggerated) below. Add arrows to the drawings to indicate the direction of the forces involved and opposite describe the effects of this reaction.

.... TORQUE REACTION

.... TORQUE REACTION

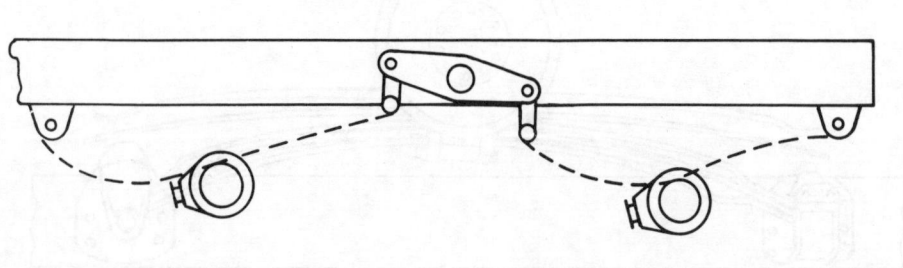

If a suspension does not counteract torque reaction and load variation occurs (as is the case with the balance-beam type) it is known as a .... suspension.

State the effects of the balance-beam action shown above.

....

....

....

....

State the meaning of the term 'non-reactive' when used to describe a multi-axle suspension.

....

....

....

....

....

....

The tandem-axle suspension shown below is a non-reactive type. The upper and lower torque arms locate the axles to resist drive and brake torque reaction, thereby preventing load variation between the axle and relieving the road springs of torque reaction stresses.

Complete the labelling on the drawing.

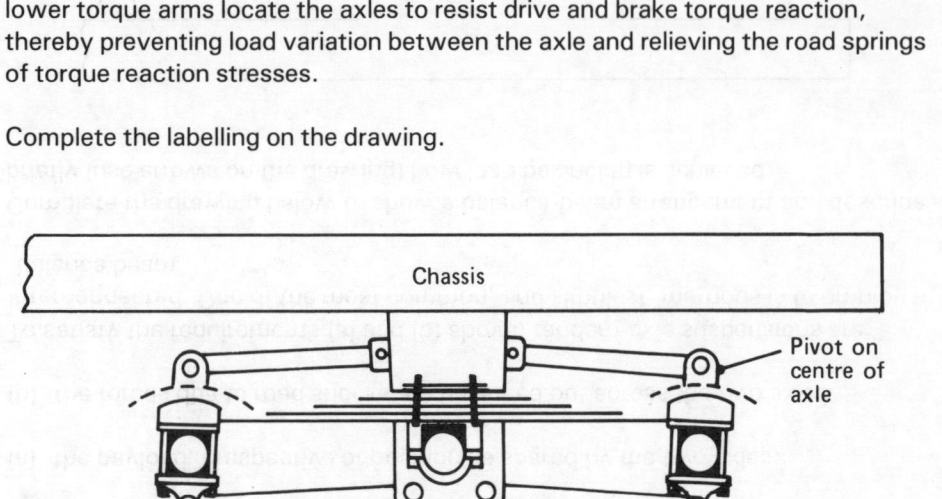

Chassis

Pivot on centre of axle

Complete the drawing below to show the bell crank lever type of non-reactive suspension and describe the action of the suspension.

....

....

....

....

....

....

AXLE LIFTS

Axle lifts are usually employed on second steered axles or dead axles in multi-axle layouts. The purpose of an axle lift is to raise the wheels of one axle clear of the ground when the vehicle is operating unladen. The resulting benefits are:

....

....

....

Make a sketch to illustrate an axle-lift arrangement and describe briefly how it operates.

Very often the operational requirements for a particular vehicle cannot be fulfilled by the stock vehicle/chassis supplied by a vehicle manufacturer.
Some companies do therefore specialise in chassis engineering in which chassis and suspensions are modified to accommodate a specific body or type of load.

Complete the sketch below to show a modified suspension for an additional axle.

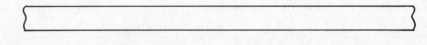

Independent suspension is employed on some PSVs, very often in conjunction with air springs. Sketch and label such a system.

Materials used for suspension components.

Complete the table below.

Component	Material	Reasons for use
Road spring		
Suspension links, radius arms, etc.		
Ball joints		
Spring eye bushes, hgvs		
Radius arm bushes		

List the preventive maintenance, MOT checks/tasks associated with suspension:

....
....
....
....
....
....
....
....
....
....
....
....
....
....
....
....
....

What benefits are to be derived from carrying out routine preventive maintenance?

....
....
....
....
....

List the general rules/precautions to be observed while carrying out routine suspension system maintenance and running adjustments:

....

....

....

....

....

....

....

....

....

....

....

....

Describe the methods of protecting the system against hazards during use or repair:

....

....

....

....

....

....

....

CHECKING SUSPENSION ALIGNMENT AND GEOMETRY

Describe briefly the suspension alignment and geometry checks normally carried out on the rear suspension: use simple sketches on the vehicle outline to illustrate how equipment is used for this purpose.

....

....

....

....

....

....

....

....

....

....

....

121

SUSPENSION SYSTEM FAULTS, SYMPTOMS, PROBABLE CAUSES AND CORRECTIVE ACTION

Complete the table in respect of the faults listed.

FAULTS	SYMPTOMS	PROBABLE CAUSES	CORRECTIVE ACTION
Abnormal excessive tyre wear			
Broken road spring			
Broken centre bolt			
Worn dampers			
Corroded spring mounting			
Worn radius arm bushes			
Worn ball joints			
Leaking spring displacers			
Missing bump stop			

25.3 FORCES DUE TO TORQUE REACTION

Calculations

The two forces acting on the spring eyes (where an axle is located in the centre of the spring) are said to form a 'couple', that is, they are equal and opposite forces acting on either side of a pivot. The torque exerted by a couple is found by multiplying one of the forces by the perpendicular distance between the forces, for example

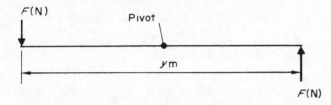

Turning moment of torque $= F \times y$ (N m)

similarly $F = \dfrac{\text{torque}}{y}$

Problem

1. A leaf spring centrally attached to a driving axle measures 1.4 m between spring eye centres. If the driving torque in the half shaft is 1000 N m, calculate the thrust on each spring eye as a result of drive torque reaction.

$$\text{Upward thrust on front eye of spring} = \frac{\text{torque}}{\text{distance between spring eyes}}$$

$$\text{upward thrust on front eye of spring} = \frac{1\,000}{1.4} = 714.28 \text{ N } Ans.$$

$$\therefore \text{downward thrust on rear eye of spring} = 714.28 \text{ N } Ans.$$

2. Consider the spring below, normally loaded as shown. Calculate the load supported by the front and rear spring eyes respectively when the driving torque in the half shaft is 1200 N m.

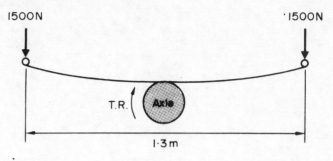

....
....
....

3. If the load on the front eye of a rear spring decreases by 600 N and the load on the rear eye of the spring increases by the same amount during braking, calculate the braking torque if the spring measures 1.2 m between the eyes.

....
....
....

4. During acceleration the torque in the propeller shaft of a vehicle is 1280 N m. If the rear axle ratio is 4.5 : 1, determine the change in load on each of the spring eyes, which are 1.5 m between centres, owing to torque reaction.

....
....
....

Chapter 7

Steering

STEERING

The steering system provides a means of changing or maintaining the direction of a vehicle in a controlled manner. The functional requirements of the system are:

1. *Driver effort should be minimal*
....

2. ....
....

3. ....

LAYOUTS

Beam axle, single track rod (light and hgvs)

Add the steering linkage to the layout below to illustrate a beam axle steering system.

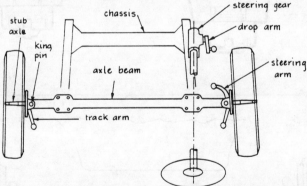

With this system, as the axle tilts owing to suspension movement, the single track rod moves correspondingly and the steering is unaffected. It is unacceptable however to employ a single track rod linkage with independent front suspension (ifs). Why is this?

....
....
....
....

Divided track rod

Car steering system

Virtually all cars and many light goods vehicles have independent front suspension (ifs). The steering linkage must therefore be designed to accommodate the up and down movement of a steered wheel without affecting the other steered wheel. In most systems two short track rods which pivot in a similar arc to the suspension links are connected to the stub axles.

One system in use can be described as: a steering gearbox with 'idler' and three-track-rod layout.

Complete the drawing of such a system shown below by adding the linkage and labelling the parts.

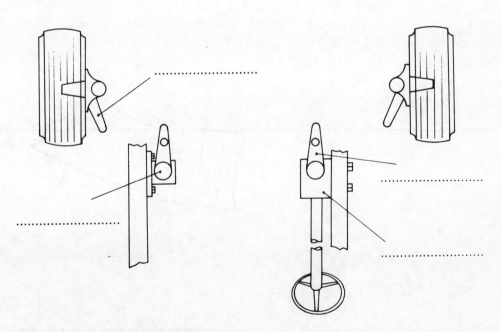

What are two main disadvantages of this layout?

....

Rack and pinion

Examine a vehicle fitted with a rack-and-pinion steering system and complete the drawing below to illustrate this; label all the parts.

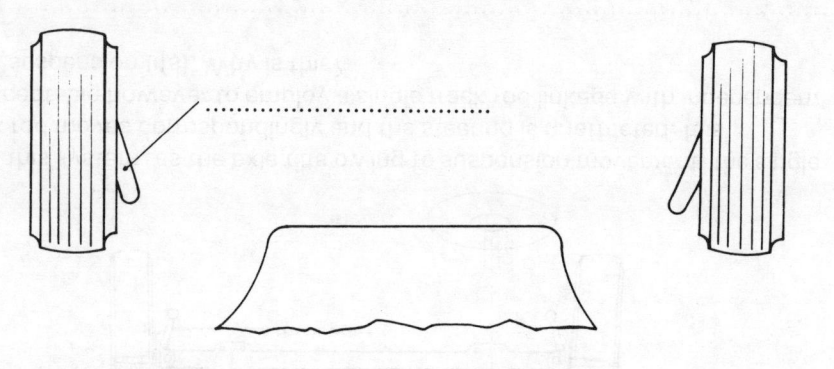

Linkage ball joints

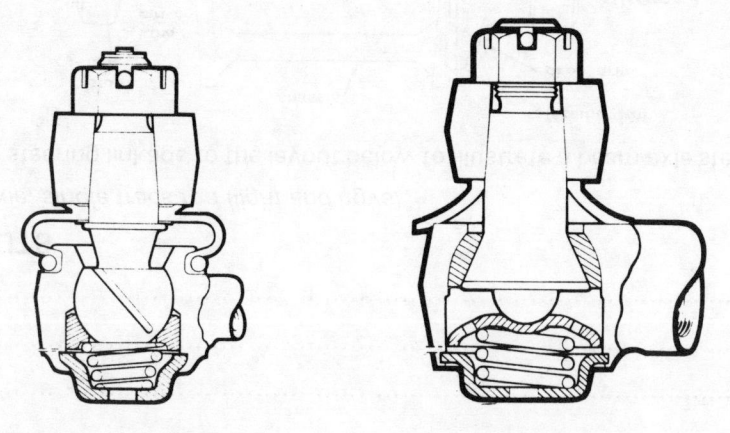

The joints shown incorporate a ball-pin and spring loaded bearing socket. State the purpose of the spring.

....

....

....

Steering swivel assembly (stub axle)

Examine a beam axle steering system and complete the drawing below to show the king pin and stub axle assembly.

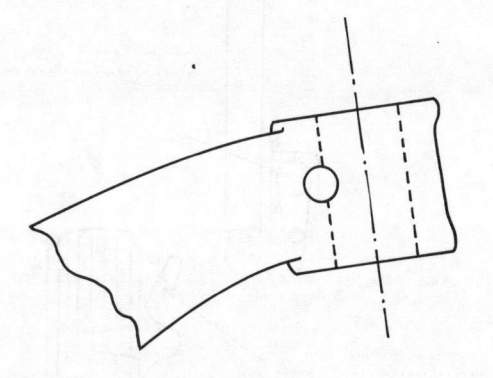

State the purpose of the king pin thrust bearing:

....

....

How is the king-pin located in the axle beam?

....

....

How are the king-pin bushes lubricated?

....

26.5, 48.5, 49.5 Describe the steering action in the suspension systems shown below. Indicate the steering pivots on the drawings.

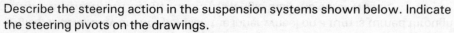

Double wishbone

....
....
....
....
....
....
....
....
....
....
....

Macpherson strut

....
....
....
....
....
....
....
....
....

FRONT HUB

INVESTIGATION

Examine a front hub and complete the drawing below to include bearings and grease seal. Label all parts.

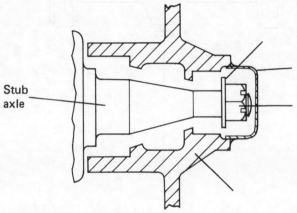

Stub axle

What type of bearings are used?

....

Name two other types of hub bearing:

1. .... 2. ....

A popular type of grease seal used in front hubs is the spring loaded rubber seal. Name two other types of grease seal:

1. .... 2. ....

State three reasons why grease can leak past a serviceable seal:

1. ....

2. ....

3. ....

127

26.3 TRUE ROLLING

When a vehicle travels on a curved path during cornering, true rolling is obtained only when the wheels roll on arcs which have a 'common centre' or common axis. Show the common centre on the drawing below.

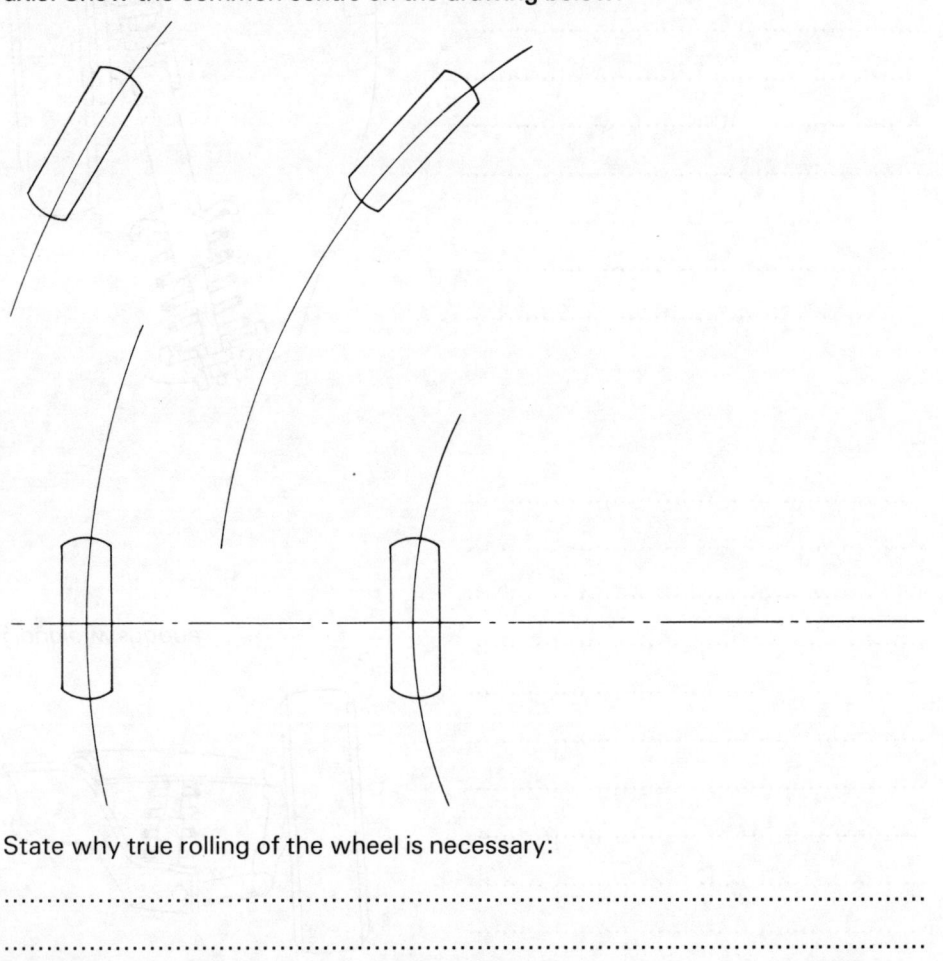

State why true rolling of the wheel is necessary:

....

....

To achieve true rolling while cornering, the wheels are steered through different angles; that is, they are not parallel; the inner wheel on a turn is turned through a .... angle than the outer wheel.

ACKERMAN SYSTEM

The difference in steering lock angles to give true rolling of the wheels while cornering can be achieved by making the track rod shorter than the distance between the king-pin centres; that is, the steering arms are usually inclined inwards. If the track rod is in front of the axle, as on many cars, then it is longer than the distance between the king-pin centres and the steering arms must be inclined outwards.

Add a track rod and steering arms to the drawing below and extend lines through the steering arms to show the point of intersection on the vehicle centre line if the linkage is to satisfy conditions for true rolling.

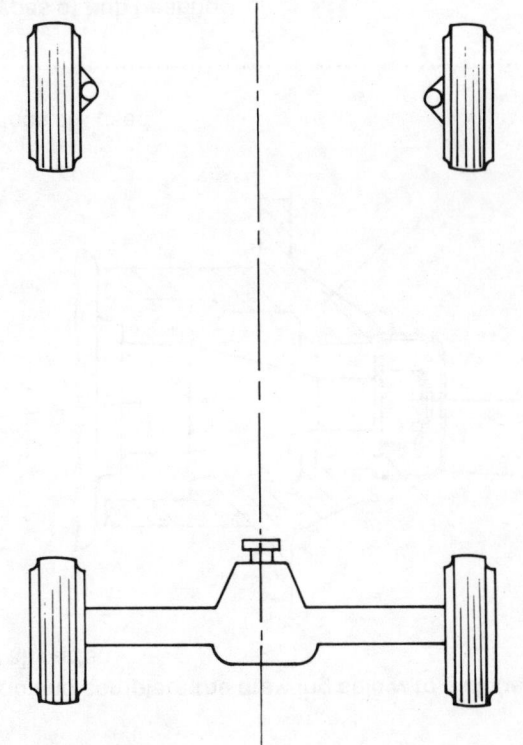

This type of steering linkage is known as:

....

The diagram below shows the position taken up by the track rod when the wheels are on lock. Indicate on the drawing below a typical angle for the inner wheel for the angle of lock shown on the outer (left-hand) wheel.

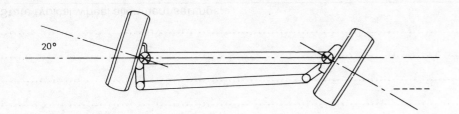

20°

The difference in lock angles shown above is known as

....

Ackerman as applied to heavy goods vehicles

Using twin steering axles spreads the load and reduces axle loading.

To obtain true rolling of the steered wheels on a twin-steering axle arranement, the first axle wheels are steered through a greater angle of lock than the second axle wheels. If a tandem-axle arrangement is used at the rear, tyre scrub is inevitable owing to the fact that the rear wheels can never roll about a common centre.

To obtain as near true rolling as possible, the 'common centre' lies on a line extended from a point midway between the axles.

State how the twin-steer linkage shown at A opposite produces true rolling and sketch an alternative linkage at B.

....

....

....

....

Add dotted lines to the drawing below to show the steering geometry to give (as nearly as possible) true rolling for the eight-wheel vehicle layout shown.

State how the rear-axle spread (that is, their distance apart) affects tyre scrub.

....

....

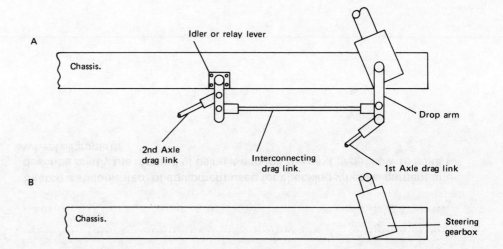

A

Chassis.

Idler or relay lever

Drop arm

2nd Axle drag link

Interconnecting drag link.

1st Axle drag link

B

Chassis.

Steering gearbox

STEERING GEOMETRY

The road wheels of a vehicle should always rotate with a true rolling motion (no tyre scrub). To ensure true rolling the steered wheels must be correctly aligned. This usually entails adjusting the effective length of the track rod to give a static 'toe-in' or 'toe-out' of the steered wheels.

The wheels shown below are .... , that is, dimension A is slightly less than dimension B.

When dimension B is slightly less than dimension A the wheels would be said to have ....

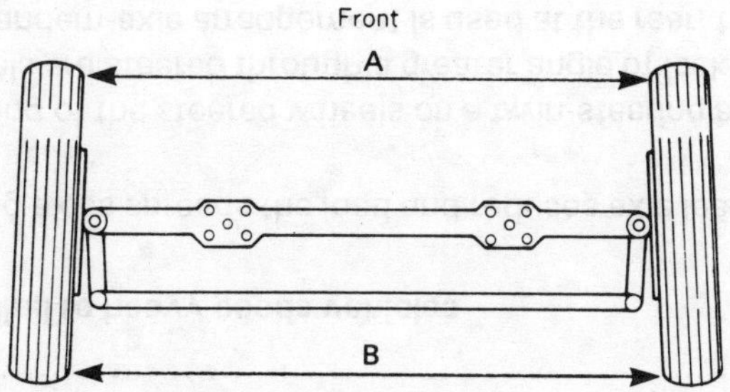

State the purpose of toe-in or toe-out:

....
....
....
....
....

State typical wheel alignment settings:

Vehicle make	Model	Toe-in	Toe-out

State the main effects of incorrect wheel alignment:

....
....
....

Sketch a suitable item of equipment used for checking wheel alignment and describe briefly the method of using it and (on the next page) how to adjust wheel alignment.

Checking wheel alignment

....
....
....
....
....
....

Adjusting wheel alignment

26.8, 49.9

....
....
....
....
....
....
....
....

TWIN STEER ALIGNMENT

One type of optical alignment gauge is shown below in position to check the rear axles for parallelism. This is part of the wheel alignment check.
Each front axle is checked individually for alignment as in single axle steering.

Describe the further checks necessary with this steering arrangement:

....
....
....
....
....
....
....
....
....
....
....
....
....
....
....
....

Illustration 5

131

Toe-out on turn

The inner steered wheel turns through a greater angle than the outer wheel when cornering (Ackerman) to give an amount of 'toe-out on turn'.

Describe a method of checking and adjusting toe-out on turn, and state a typical setting for this.

....

....

....

....

Lock stops

Adjustable lock stops in the steering mechanism limit maximum lock to avoid tyres fouling chassis and prevent extreme travel of the steering gear. Examine a car or hgv and sketch the lock stop arrangement.

CASTOR ANGLE

One of the desirable features of a steering system is the ability of the road wheels to self-centre after turning a corner. This self-centring effect can be achieved by designing the wheel and steering swivel assembly so that the wheel centre trails behind the swivel axis.

Make a simple sketch on the right below to show how the castor is applied to vehicle; indicate the castor angle.

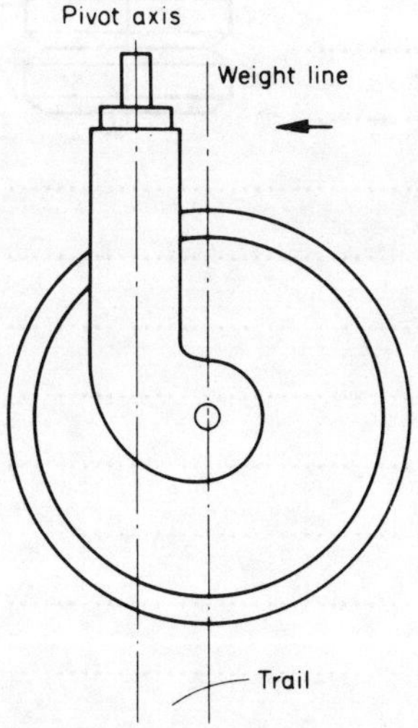

Pivot axis
Weight line
Trail

Principle of castor action   Castor applied to vehicle

Complete the table below by adding typical castor angles.

Car	Light commercial vehicle	Heavy commercial vehicle

CAMBER ANGLE AND SWIVEL PIN INCLINATION

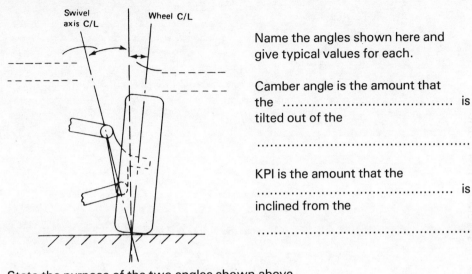

Name the angles shown here and give typical values for each.

Camber angle is the amount that the .... is tilted out of the

....

KPI is the amount that the

.... is inclined from the

....

State the purpose of the two angles shown above.

Camber

....
....
....
....

KPI

....
....
....
....
....

Centre-point steering

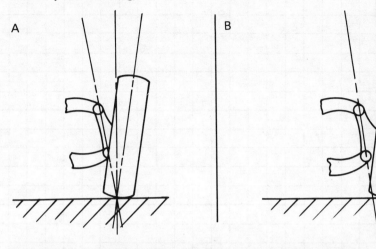

The drawing at A above illustrates true centre-point steering (the wheel centre line and swivel axis centre line converge at ground level). Many vehicles, however, adopt the arrangement shown at B; this shows the wheel and swivel axis centre lines converging below ground level to give a 'positive offset'; this give a better steering action.

Negative offset

Complete the drawing below by adding a swivel assembly to show a 'negative offset' and state the reasons for this arrangement.

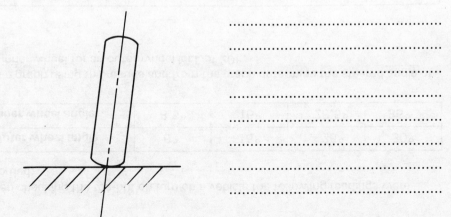

....
....
....
....
....
....
....
....

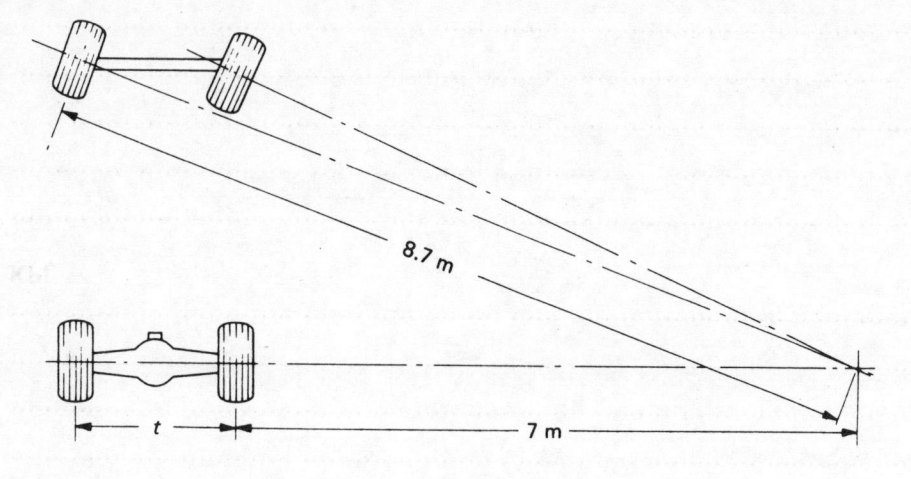

The diagram above illustrates the conditions necessary for true rolling when a vehicle is cornering. Determine the track measurement of the rear wheels (*t*) for the vehicle shown if the wheelbase is 2.5 m, and the steered angles for both steered wheels.

When checking the toe-out on turn on a vehicle the following readings were recorded:

Outer wheel angle	8°	15°	25°	30°
Inner wheel angle	8.5°	16°	28.5°	35°

Plot a graph using the results and from the graph determine the steered angle of the inner wheel for an outer wheel lock of 20°.

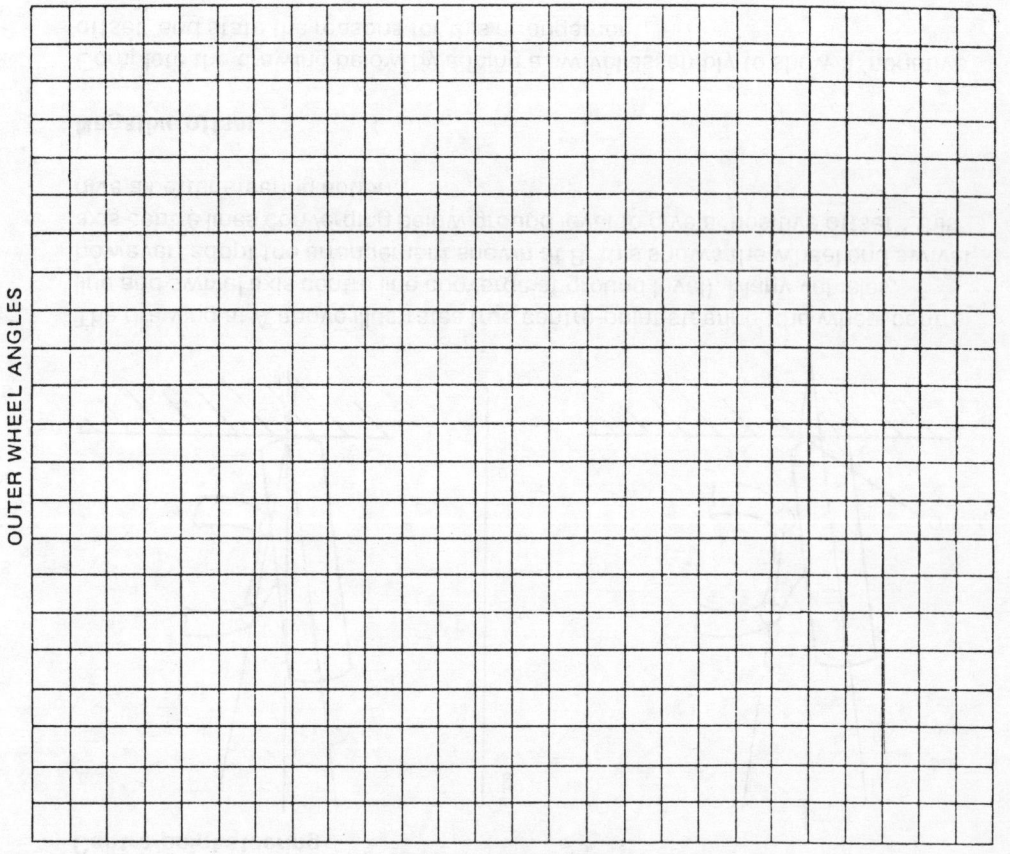

OUTER WHEEL ANGLES

INNER WHEEL ANGLES

26.5 STEERING GEAR

The steering gearbox is incorporated into the steering mechanism for two main reasons:

(a) To change the rotary motion of the steering wheel into the 'to and fro' movement of the drag link or rack.

(b) ....

....

....

A number of different types of steering gears are in use; complete the list below to name five types.

1. Cam and peg.

2. ....

3. ....

4. ....

5. ....

State the desirable characteristics of steering gearboxes:

....

....

....

....

....

The cam and peg steering box has been superseded by the recirculating ball, worm and roller, and rack and pinion gear, the latter being by far the most popular for car applications.

Complete the drawing below to show a worm-and-sector type steering gear.

Worm-and-sector steering box

State why the worm is 'hour-glass'-shaped:

....

....

State the purpose of the bearings and name the type of bearing used in the steering box shown:

....

....

....

....

....

How is the running clearance of the bearings adjusted?

....

....

State one disadvantage of this form of steering gear:

....

....

WORM AND ROLLER STEERING BOX

In what way is the worm and roller an improvement on the cam and peg steering gear?

....

....

....

....

Add the cross shaft assembly to the drawing at (b) and include the mesh adjuster.

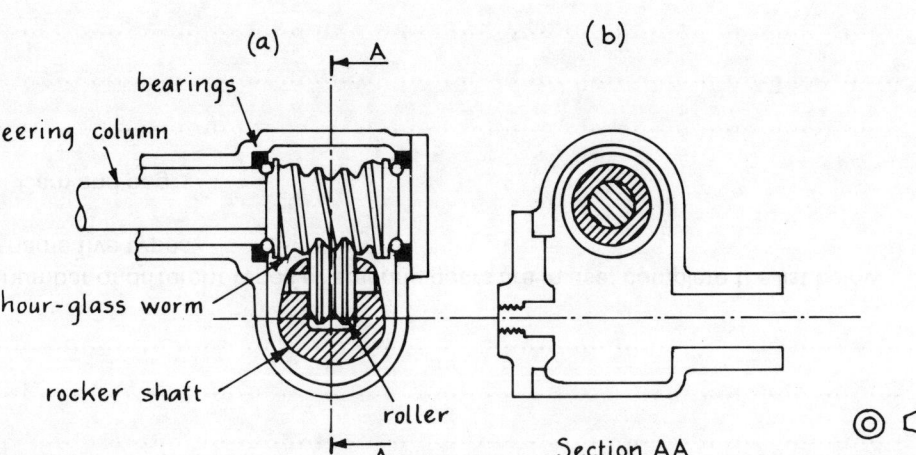

Name two vehicles, a car and a goods vehicle, which use this type of gear:

....

RECIRCULATING-BALL STEERING BOX

The recirculating-ball steering box is a development of the worm and nut type. A row of ball bearings operate in grooves between the nut and worm to form in effect a screw thread; in this way, rolling contact between the nut and worm replaces sliding contact.

Complete the simple drawing below to include the ball bearings, nut, return passage and rocker shaft.

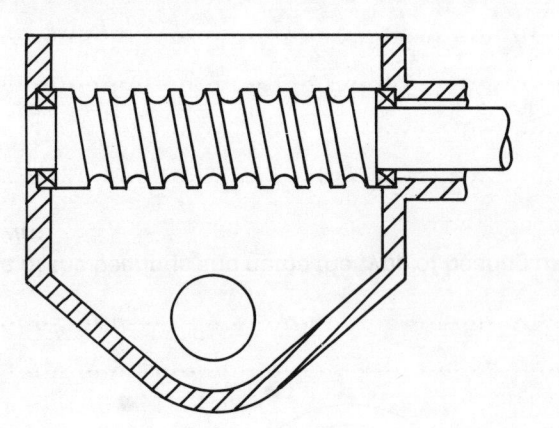

State the purpose of the return passage:

....

....

....

....

....

Name two vehicles, a car and a goods vehicle, which use this type of gear:

....

....

RACK AND PINION

Label the rack and pinion gear shown below and add the rack to the simple drawing to illustrate the action of this form of gear.

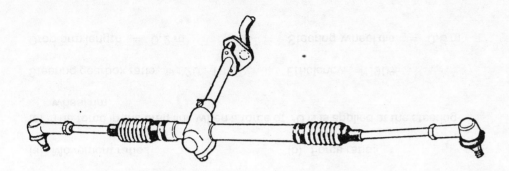

─────────────────── ·─· ─·─·─·─ rack centre line

Why is this form of steering gear being used in preference to the steering gearbox on most cars?

....

....

....

....

State typical gear ratios for:

Car steering gear ....

hgv steering gear ....

MOVEMENT RATIO, FORCE RATIO, EFFICIENCY

The efficiency of a steering gearbox is largely dependent on the type of gear employed. Two factors to be considered when calculating the efficiency are:

TORQUE RATIO and GEAR RATIO

Torque ratio   =

Gear ratio   =

% efficiency   = $\times \dfrac{100}{1}$

When considering force ratio and movement ratio two other factors must be taken into account:

1. ....

2. ....

Force ratio   =

Movement ratio =

% efficiency   = $\times \dfrac{100}{1}$

State the effect of efficiency on torque ratio:

....

....

137

Calculation

A worm and sector type steering gearbox requires a torque applied by the driver of 12 N m to produce a rocker shaft torque of 135 N m. Calculate the efficiency of the steering gear if the gear ratio is 14 : 1.

Given the information below for a recirculating-ball type steering mechanism, calculate:

(a) Movement ratio.

(b) Force ratio.

(c) The force in the drag link when a force of 70 N is applied at the steering-wheel rim.

Steering gearbox ratio = 20 : 1

Efficiency = 90%

Drop-arm length = 0.2 m

Steering-wheel dia. = 0.5 m

POWER-ASSISTED STEERING

To keep the steering wheel load to an acceptable level without increasing the number of turns required from lock to lock, a system of power assistance can be incorporated in the steering mechanism. This will reduce the effort required by the driver and also, owing to the smaller gear ratio required, keep the number of turns required from lock to lock to a reasonable figure.

Three systems in use are:

1. External ram type.

2. ....

3. ....

The power assistance on most vehicles is hydraulic, the system consisting basically of the following main parts:

....

....

....

....

The most popular hydraulic system used for this purpose is one which operates on the .... principle.

....

....

....

....

....

....

....

State the functional requirements of PAS:

....

....

....

What is the main difference between a Power-Assisted and a Power Steering system?

....

....

....

EXTERNAL TYPE PAS

In the external type PAS the power ram and control valve assembly are incorporated into the steering linkage.

Add the power ram, control valve and piping to the hgv system below and describe briefly the operation of the system.

hgv layout (external)

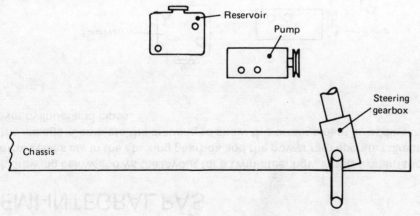

Reservoir

Pump

Steering gearbox

Chassis

Operation (external type)

....

....

....

....

....

....

Control

Add a spool valve to the simplified drawing below and describe the operation of the system.

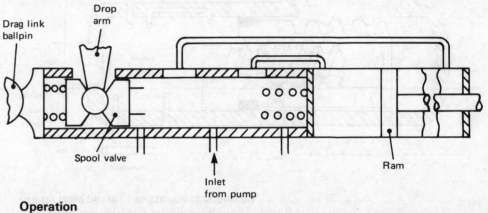

Drag link ballpin

Drop arm

Spool valve

Inlet from pump

Ram

Operation

....

....

....

....

SEMI-INTEGRAL PAS

The drawing below shows the layout for a twin-steer hgv. In this system the control valves are in the steering gearbox and the power ram operates directly on to the linkage. Complete the drawing to show this arrangement by adding the power cylinder and pipes.

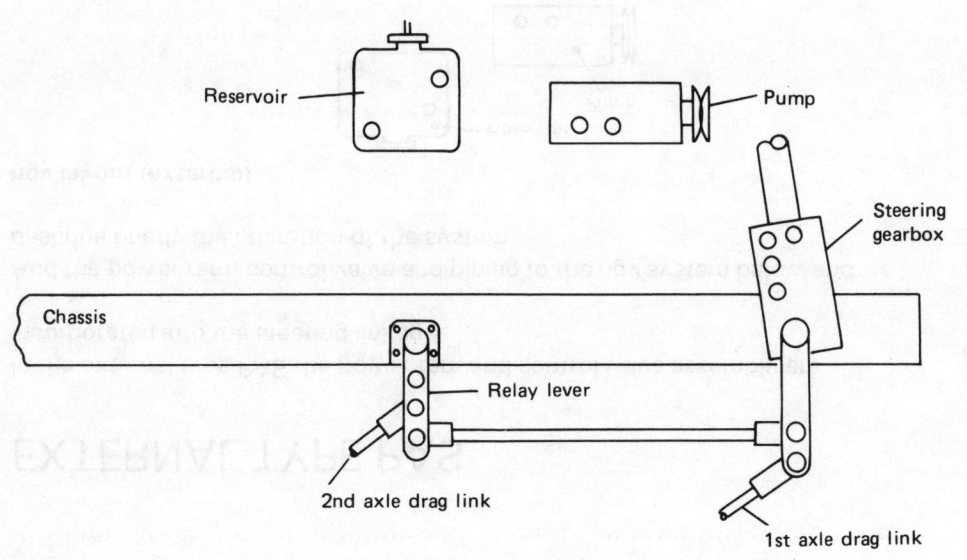

What advantages has this system over the external or integral systems?

....
....
....

Slight angular movement of the power cylinder takes place during operation. How is this accommodated?

....
....

INTEGRAL PAS

With the integral system, the servo piston or ram and control valve assembly are located in the steering gearbox or rack.

A power-assisted recirculating ball type steering box is shown below.

Complete the labelling on the simplified arrangement shown below and describe briefly how power assistance is achieved.

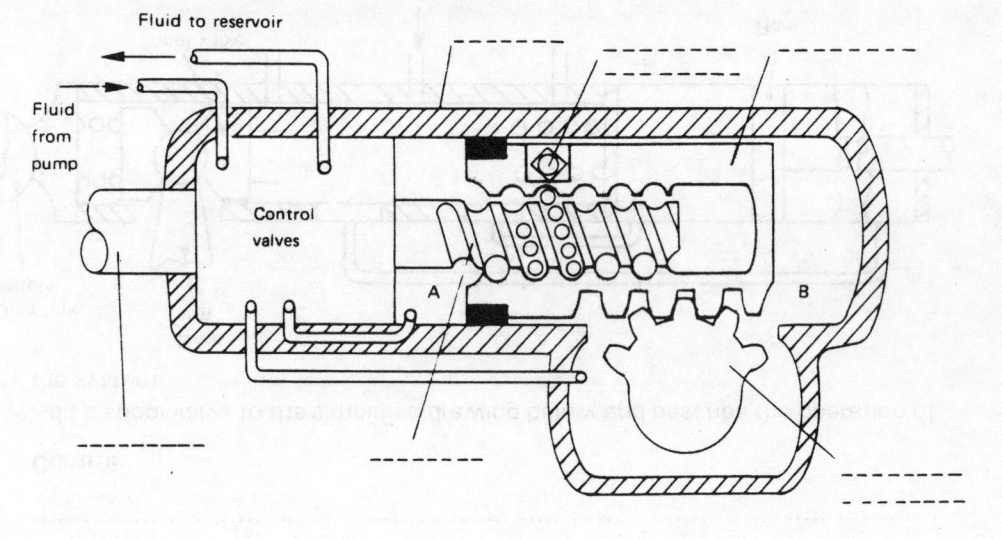

Operation

....
....
....
....
....
....

RACK AND PINION PAS

Name the main parts of the system layout shown below.

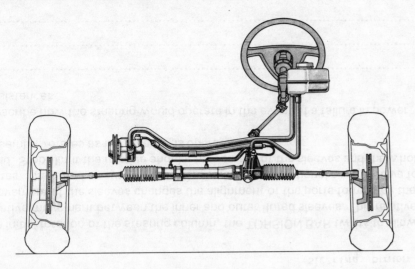

Make a simple drawing below to show how the rack is hydraulically assisted.

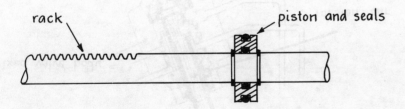

rack

piston and seals

In addition to the control valve spring or torsion bar providing sensitivity or driver feel, it is desirable to increase the required driver effort in proportion to the power assistance demanded; this gives a degree of natural feel to the system.

How is this normally achieved?

....

....

....

....

....

....

....

....

....

State the average pull required at the steering wheel rim to overcome the control valve spring pre-load on:

(a) a car ....

(b) a hgv ....

Many PAS systems, especially on cars, are described as 'speed sensitive'. What does this mean and why is it employed?

....

....

....

....

....

....

POWER-ASSISTED STEERING

8.5, 49.5

Control of power assistance

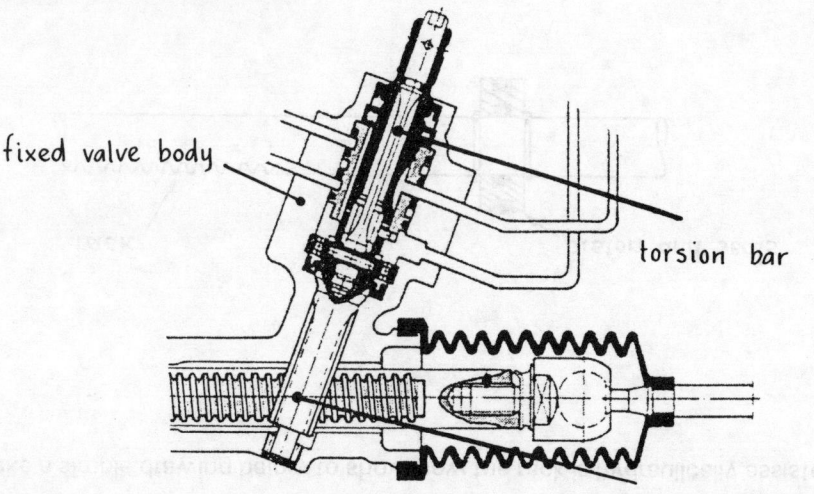

fixed valve body

torsion bar

steering pinion

On initial rotation of the steering column, the TORSION BAR twists to allow relative movement between the inner and outer fluted sleeves. This relative movement of the sleeves changes the alignment of the ports formed by the flutes. The fluted sleeves do therefore serve as a directional control valve for the fluid. Stops limit the relative angular movement of the sleeves and the whole assembly rotates as the steering is operated.

Describe how the steering would operate in the event of a failure in power assistance:

....
....
....
....
....

Valve operation

Complete drawings (b) and (c) to show the valve position and fluid direction on left and right lock.

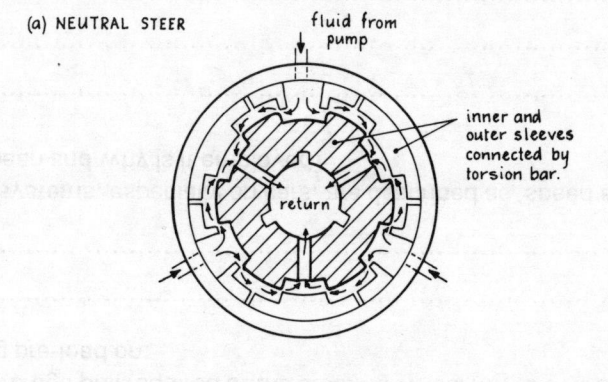

(a) NEUTRAL STEER
fluid from pump
inner and outer sleeves connected by torsion bar.
return

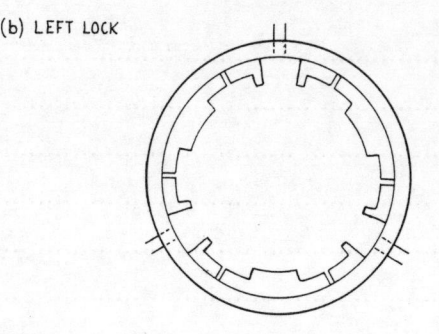

(b) LEFT LOCK

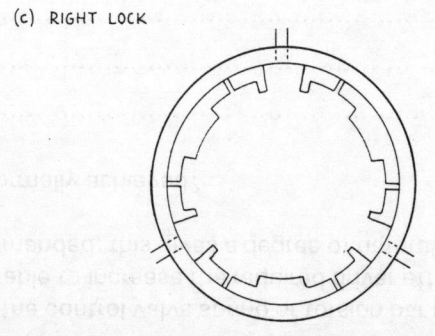

(c) RIGHT LOCK

142

Hydraulic pump and flow relief valves

The type of hydraulic pump shown below is used in many PAS systems. Name the pump, indicate on the drawing the fluid inlet and outlet points, and describe its operation. Label the drawing.

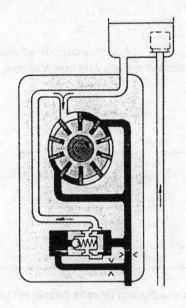

....
....
....
....
....

State the purpose and describe the operation of:

(a) the pressure relief valve and

(b) the flow control valve

Pressure relief valve

Purpose:

....
....
....
....
....

Operation:

....
....
....
....
....

Flow control valve

Purpose:

....
....
....

Operation:

....
....
....
....
....

Steering system maintenance

State the benefits of routine preventive maintenance:

....

....

....

....

....

....

How is the steering system protected against the ingress of moisture and dirt during repair and use?

....

....

....

....

....

....

List the general rules for efficiency and any special precautions to be observed when carrying out maintenance and repair.

....

....

....

....

....

List typical preventive maintenance tasks and Department of Transport (MOT) checks associated with the steering system:

....

....

....

....

....

....

....

....

....

....

....

....

....

....

....

PAS check:

....

....

....

....

....

Note: During maintenance, 'check' also means adjust if necessary.

How much free play is allowable at the steering wheel rim?

....

Describe briefly how to check steering linkage ball joints for wear:

....
....
....
....
....
....

Show on the sketch below how king-pins (or ball joints), king-pin thrust bearings and wheel bearings are checked for wear.

State the possible causes of axial movement in the steering column.

....
....
....

Give examples of the use of the following items of equipment in steering system maintenance.

TURNTABLES:

....
....

PRESSURE TEST EQUIPMENT:

....
....

BELT TENSION GAUGE:

....
....

DIAL TEST INDICATOR:

....
....

CRACK DETECTOR:

....
....

State a grade of oil suitable for use in PAS systems.

....
....

STEERING SYSTEM FAULTS, SYMPTOMS, PROBABLE CAUSES AND CORRECTIVE ACTION

Complete the table in respect of the faults listed.

FAULTS	SYMPTOMS	PROBABLE CAUSES	CORRECTIVE ACTION
Partially seized king-pins			
Worn track rod end			
Loose tie bar			
Incorrectly fitted drop arm			
PAS fluid loss			
Faulty PAS pump			
Slack PAS belt			
Excessive rocker shaft end float			
Worn or collapsed king-pin thrust bearing			

Chapter 8

Tyres and Wheels

ELEMENT 27 **UNITS 19/41/42**

TUBED AND TUBELESS TYRES

27.1/5

The modern tyre has evolved over the past 100 years from the simple pneumatic cycle tyre to the sophisticated tread and cord structures of today.

Tyres may be of a tubeless design or require the fitting of an inner tube.

Complete the drawings below to show the difference between tubed and tubeless tyres.

Tubed

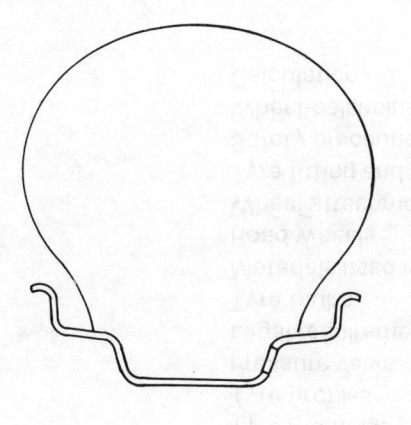

Tubeless

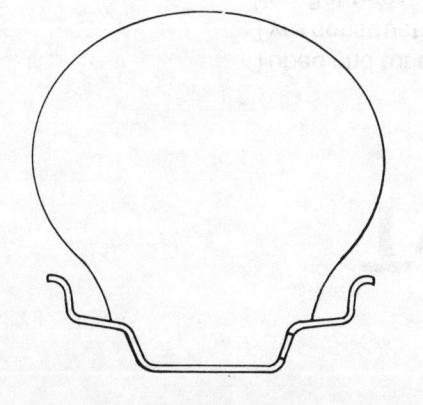

A significant feature of the tubeless tyre is the inner lining of soft rubber which extends to the outer edge of the bead. This rubber provides an effective seal at the bead and will automatically form a seal around a penetrating foreign object.

The drawing below illustrates tubed and tubeless design for truck applications.

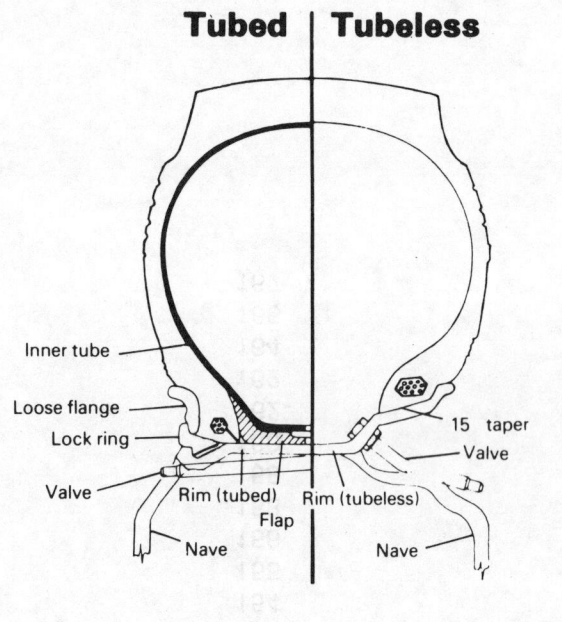

State the particular advantages of tubeless construction when employed on trucks:

....
....
....
....
....
....
....
....

148

State the purpose and functional requirements of tyres:

....

....

....

....

....

....

TYRE CONSTRUCTION

CROSS – PLY TYRES

The cords are set in layers which cross each other at about an angle of 100° and 30 to 40° to the tyre centre line.

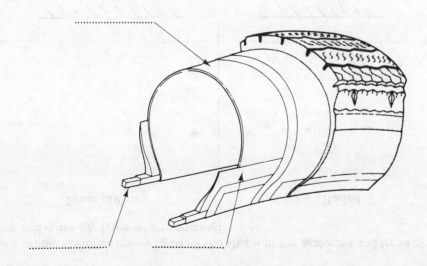

Show on the diagram the direction of the cord layers. Name the arrowed parts.

RADIAL-PLY TYRES

Arranging the cords at an angle of 90° to the bead and using breaker strips to form a circumferential belt allows a substantial reduction in the number of casing piles.

Complete the drawing below to show the direction of cords in radial construction and name the arrowed parts.

State the purpose of the breaker strips:

....

....

....

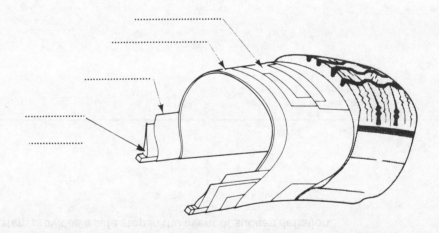

Increased tyre life is one advantage of a radial-ply tyre when compared with a cross-ply. State the main reasons for this reduction in tyre-wear rate:

....

....

....

....

The relatively rigid casing of a cross-ply tyre creates a reduction in tread contact area when the tyre is subjected to a side force during cornering.

Make simple sketches below to show the shape of the sidewalls and tread of radial and cross-ply tyres when cornering.

Cross-ply   **Radial**

State why cross-ply tyres are not as suitable as radial-ply tyres for sustained high-speed operation:

....
....
....
....
....
....

How does the structure of BIAS BELTED tyres compare with radial or cross-ply structure?

....
....
....
....

RUN-FLAT TYRES

As the name implies, the run-flat tyre is a tyre that remains operational in the event of a puncture. When a conventional tyre is deflated, continued running on the tyre will cause it to slide on the rim with the bead entering the fitting well and the rim digging into the road surface. If a tyre bead is dislodged following deflation, safe steering control is lost and severe damage to the tyre occurs.

Examine a TD type tyre and rim and describe with the aid of a sketch how the system provides a safe stop in the event of sudden deflation.

....
....
....

An interesting run-flat development is the CONTI TYRE SYSTEM (CTS). In this system the tyre sits internally on the wheel rim.

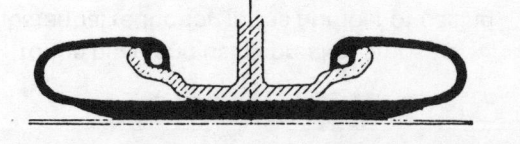

Conti Tyre System

TYRE TREAD

The tread pattern is designed to suit the particular application for the tyre.

State the tread design features for:

Steerability

....

Traction

....

Water dispersal

....

State the operational suitability for the two car tread patterns shown and sketch on the outlines opposite TWO different hgv tread patterns, stating the use in each case.

....
....
....

....
....
....

....

....

Give examples of tyre faults associated with the following symptoms:

Heavy steering

....
....

Vibration

....

Abnormal noise

....
....

Poor ride

....

Poor traction

....
....

TYPICAL TREAD PATTERNS AND SIDEWALL MARKINGS

Certain high performance tyres have an ASYMMETRIC tread pattern. State the purpose of this tread design and show by sketching at (a) an asymmetric tread pattern.

(a)

....
....
....
....
....
....
....

Each tyre manufacturer uses a coding system relating to tread design and compounds.

Shown opposite is part of the system used by the Michelin Tyre Company. Letters indicate particular tyre applications and are included, along with other information, in the sidewall markings.

Make a simple sketch to illustrate a TREAD DEPTH OR WEAR INDICATOR (built into the tyre) and state its purpose.

MXV
High speed, high performance radial with outstanding braking, precision handling and excellent grip. Rapid water dispersal to improve adhesion in wet conditions.

MX
Ideal for the family car. Excellent grip in the wet, more miles per gallon and long service.

XM+S200
Radial winter tyre designed for "Arctic" conditions, giving extra grip from special rubber compounds that withstand extremely low temperatures. Wide grooves and reinforced shoulders break up ice and snow and make the tyre particularly effective on muddy or other yielding surfaces. Can be used all year round.

MXX
A radial tyre for ultra high performance sport cars giving an extremely high level of grip, braking power, high torque transmission and reliability.

LOAD INDEX

A 'load index' table is shown below. State its purpose.

LT	kg	LT	kg	LT	kg	LT	kg	LT	kg	LT	kg
0	45	20	80	40	140	60	250	80	450	100	800
1	46.2	21	82.5	41	145	61	257	81	462	101	825
2	47.5	22	85	42	150	62	265	82	475	102	850
3	48.7	23	87.5	43	155	63	272	83	487	103	875
4	50	24	90	44	160	64	280	84	500	104	900
5	51.5	25	92.5	45	165	65	290	85	515	105	925
6	53	26	95	46	170	66	300	86	530	106	950
7	54.5	27	97.5	47	175	67	307	87	545	107	975
8	56	28	100	48	180	68	315	88	560	108	1000
9	58	29	103	49	185	69	325	89	580	109	1030
10	60	30	106	50	190	70	335	90	600	110	1060
11	61.5	31	109	51	195	71	345	91	615	111	1090
12	63	32	112	52	200	72	355	92	630	112	1120
13	65	33	115	53	206	73	365	93	650	113	1150
14	67	34	118	54	212	74	375	94	670	114	1180
15	69	35	121	55	218	75	387	95	690	115	1215
16	71	36	125	56	224	76	400	96	710	116	1250
17	73	37	128	57	230	77	412	97	730	117	1285
18	75	38	132	58	236	78	425	98	750	118	1320
19	77.5	39	136	59	243	79	437	99	775	119	1360

....
....
....

....

The table top right shows 'speed ratings' for tyres. These are maximum safe speeds of operation.

State the relationship between LOAD INDEX and SPEED RATING:

....
....
....
....

SPEED RATING

Speed symbol	Speed (km/h)
L	120
M	130
N	140
P	150
Q	160
R	170
S	180
T	190
U	200
H	210
V	240

TYRE SIZES AND RATINGS

The tyre size is marked on the .... of the tyre.

Older tyres are marked with two dimensions.

The first states the ....

The second the ....

During the 1960s, radial section widths were stated in .... and cross-ply in ....

Most manufacturers still adopt this method of size marking (Metric/Imperial).
One exception to this is in the marking of certain tyres which are designed specifically for a particular rim.
In this case the rim diameter is given in millimetres, such as Michelin TDX tyres.
Give THREE examples of different types of tyre sizes.

Tyre size ....   Make ....   Application ....

....

....

TYRE PROFILES

Before about the early 1950s, tyres were designed to a basic circular cross-section — the tyre width being approximately equal to the radial height.

Development in racing clearly demonstrated that a tyre with a low profile gives many advantages.

This wider, lower type of tyre is said to have either a low height/width, low profile or low aspect ratio.

The height to width is quoted as a percentage, such as 70%; for a tyre with a 70% aspect ratio, therefore, the height dimension is 70% of the width.

The aspect ratio is normally indicated on the sidewall marking, if however it is not indicated, it usually means that the tyre is of the now normal aspect ratio which is about 82%.

Tyre having 100% profile ratio

Draw a tyre having a 70% profile ratio

Racing-car tyre 35% profile ratio

State the advantages of low-profile tyres when compared with 100% profile tyres.

....

....

....

....

A tyre marking of 165/70 R 13 MXL 79S means:

165 = ...

70 = ...

R = ...

13 = ...

MXL = ...

79 = ...

S = ...

Examine two cars and two hgvs and complete the table below.

Vehicle make/model	Tyre marking	Specified pressure	Tubed/ tubeless	Radial/ cross-ply

On certain tyres, load-carrying capacity is shown as a PR number, such as

car: PR 4. Truck: PR 14.

PR means ....

The manufacturer's specifications for original equipment is the best guide when choosing tyres for a vehicle. Deviations from this however may be when a vehicle is used for particular applications, such as rallying, cross-country operation, travelling in different climatic conditions abroad etc.

PRESSURE VALVE

All types of penumatic tyres must be fitted with a valve in order to retain the air pressure. The valve is of a similar construction in both tube and tubeless tyres. Inspect a tyre valve similar to the one opposite and complete the assembly, naming the main parts.

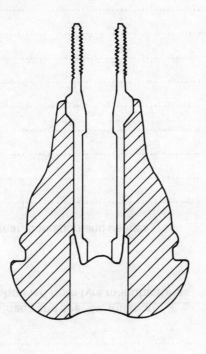

....

....

....

....

....

What is the object of fitting a pressure cap?

....

....

How do tyre pressures relate to vehicle LOADING and HANDLING?

....

....

....

....

Examine a dismantled hgv wheel, tyre and inner tube assembly and make sketches to show:

1. The valve stem arrangement.

2. The method used to protect the inner tube from chafing against the wheel rim.

How do the tyre pressures vary from, say, the pressure before the start of a long, fast journey to the pressure immediately after, and what is the reason for this variation?

....

....

If sustained high-speed driving is to be undertaken, what adjustment should be made to the tyre pressures?

....

....

LEGAL REQUIREMENTS

The tyres fitted to a vehicle must conform to the current legal requirements as laid down in the construction and use regulations.

Tyres selected for a vehicle must be the correct size and type in relation to:

1. The wheels.

2. ....

3. ....

4. ....

Information relating to: rim fitment, maximum loads and speeds, inflation pressures and minimum dual spacing, can be obtained from tyre manufacturers' tables.

State the legal requirements with respect to wear, condition and usage:

....

....

....

....

....

....

....

....

....

....

....

Recut tyres

Tread recutting or regrooving is a process carried out using an electrically heated blade to restore a tread pattern as it approaches illegality, thereby increasing the tyre life.

There are two different recutting processes:

1. The original tread depth is increased by cutting into the base rubber between the lowest point of the original grooves and the casing plies.

2. ....

....

....

....

To what depth is recutting normally carried out?

....

....

....

....

....

On which vehicles can re-cut tyres be legally used?

....

....

....

....

....

27.4 TYRE COMBINATIONS

Various legal and illegal combinations of radial- and cross-ply tyres are shown on this page. Write legal or illegal, as appropriate, under each combination.

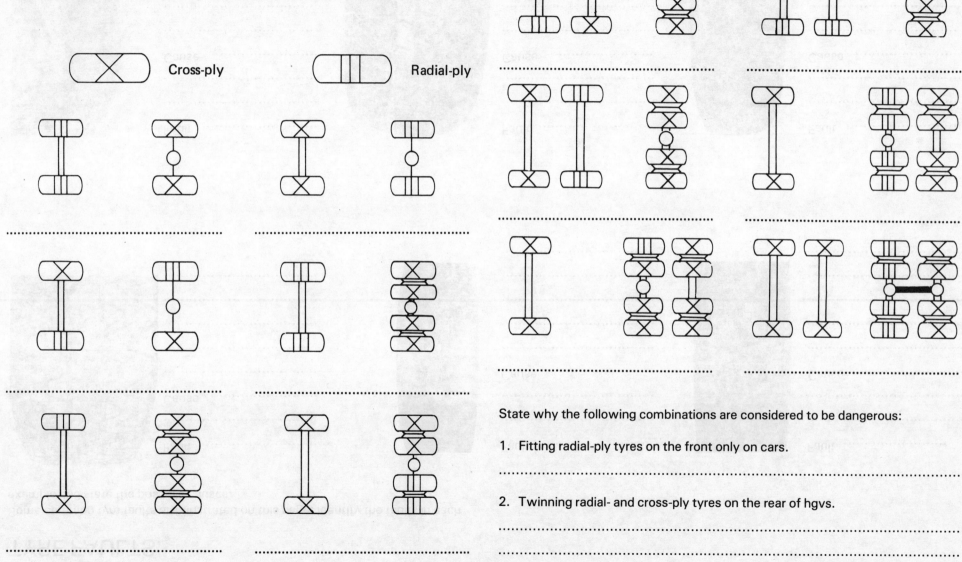

State why the following combinations are considered to be dangerous:

1. Fitting radial-ply tyres on the front only on cars.

2. Twinning radial- and cross-ply tyres on the rear of hgvs.

157

TYRE FAULTS

Some common tyre faults are illustrated on this page; identify the faults in each example and state the possible causes.

Fault ....

....

Cause ....

....

....

....

Fault ....

....

Cause ....

....

....

....

....

Fault ....

....

Cause ....

....

....

....

....

Fault ....

....

Cause ....

....

....

....

....

Fault ....

....

Cause ....

....

....

....

....

....

Fault ....

....

Cause ....

....

....

....

....

MATERIALS USED IN TYRE AND WHEEL CONSTRUCTION

Casing

This is the body of the cover (that is, tyre) and is built up of a number of layers or plies of rubber-coated cord. Each layer of material does not have threads running in two directions, but all the main threads run in one direction. This is to avoid the chafing that would occur with the normal weft and warp type of fabric.

Name the two most widely used tyre cord materials:

1. ....   2. ....

What material is used to reinforce the tyre bead?

....

Where is natural rubber used in a tyre?

....

The ply material has its main threads all running in one direction. This means that those threads will very easily separate if the material is pulled sideways. How is this overcome (a) on cross-ply tyres and (b) on radial-ply tyres?

(a) Cross-ply

....

....

....

(b) Radial-ply

....

....

....

The compound used for the tread is a mix of natural and synthetic rubber in which additives are included to improve its properties of grip, wear and response. The most essential single additive is carbon black which helps to give the tyre tread resistance to wear.

Name any other important tread rubber additives and state their function:

....

....

....

....

What features must tyre treads embody in order to reduce the following?

Squeal ....

....

....

Pattern noise ....

....

....

....

The material used in the construction of most car and hgv wheels is ....
.... , which is strong enough to resist the torsional forces and bending loads imposed on the wheel. Name one other material used in the construction of vehicle wheels and give reasons for its use.

....

....

....

....

ROAD WHEELS

State the main purposes and functional requirements of road wheels:

....
....
....
....
....
....
....

WHEEL-RIM CONSTRUCTION

The type of wheel rim shown below is fitted to most cars and light vans; name the rim design and state the purpose of the 'well' in the centre.

....

During inflation the tyre bead is forced up the taper on the rim until it contacts the rim flange; this action provides an air tight joint between tyre and rim.

Apart from being heavier and stronger than car wheels, commercial vehicle wheels are designed to allow the fitting and removal of tyres with very rigid sidewalls and beads; many wheels are two-, three- or four-piece structures.

Semi-drop centre

The semi-drop-centre rim is a two-piece rim and is suitable for light commercial vehicles.

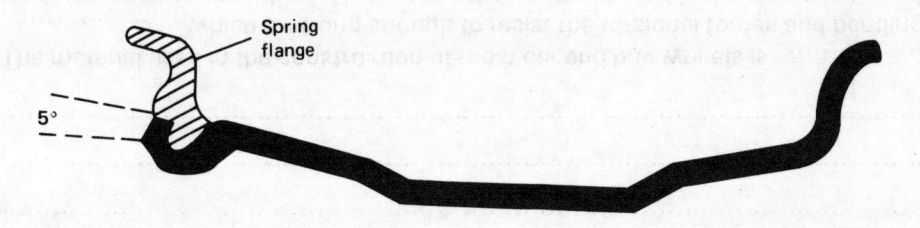

Drop centre (tubeless tyres)

Describe, with a sketch, the drop centre rim and give typical vehicle applications of this type of wheel:

....
....
....
....
....
....

Wide base

Wide-base rims have a 5° taper seat; the term 'wide base' denotes that, for a given size of tyre, the width between the rim flanges is greater than that of a 'flat-base' rim.

Two-piece

Spring flange

5°

Sketch a three-piece wide-base rim.

Wide base 3B type

The 3B type has a 3° taper seat and can be used for tubeless tyres; make a sketch to show this type:

Divided-type rim

Divided or 'split' rims are used on many military vehicles and certain commercial vehicles where large, single, heavy-duty cross-country type tyres are fitted. Tyre fitting and removal are carried out by dividing the two halves of the wheel which are bolted together.

Sketch this type of rim and outline the safety precautions associated with its use.

Note: Some very small wheels, e.g. scooter sizes, may be of the divided-rim type.

Safety (divided rims):

....

....

....

....

....

....

....

WHEEL ATTACHMENT AND LOCATION

27.1/5/7/8

The road wheel assembly is usually secured to the hub flange by studs and wheel nuts or by set bolts. Radial location of the wheel can be achieved by providing a conical seating on the wheel and using taper faced nuts, bolts or taper washers (see Level 1 Book, page 35).

An alternative method of wheel location used on cars and hgvs is to employ SPIGOT MOUNTED wheels.

Examine a vehicle using this form of wheel location and sketch and describe the system below.

Give an example and reason for the use of left-hand screw threads for wheel attachment:

....

....

....

State the purpose of and describe, briefly, SPACESAVER wheel assemblies:

....

....

....

....

....

Name two vehicles using spacesaver spare wheels:

Make .... Model ....

....

Repair and maintenance

Routine tyre maintenance helps to ensure safe, efficient operation and maintains performance and tyre life. List the routine checks and adjustments for tyres and wheels:

....

....

....

....

....

....

....

....

....

....

Tyre fitting tools and equipment range from simple tyre levers to expensive purpose-built machines. However, the basic technique is the same irrespective of the equipment being used. Complete the sketch below to show how the type of wheel shown facilitates tyre removal.

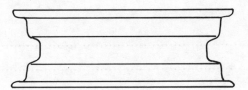

Over which rim flange (a or b) shown below would the tyre be removed and refitted and why?

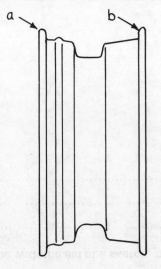

....

....

....

....

....

....

For what purpose is a bead-breaker used?

....

....

....

In addition to safety precautions, list some important points to observe when tyre fitting:

....

....

....

....

....

Describe permissible tyre and inner tube repairs:

....

....

....

....

....

....

....

....

Describe with the aid of a sketch, how tyre 'tread depth' is measured:

....

....

....

State the purpose of 'tyre rotation' and show, by sketching below, a typical rotation pattern:

....

....

....

....

Safety precautions

There are many hazards associated with tyre servicing;it is essential therefore to be aware of these and to observe certain personal safety precautions when removing a wheel from the vehicle, and removing, refitting and inflating a tyre.

List the safety points for servicing, removing and fitting tyres and wheels:

....

....

....

....

....

....

....

....

....

....

....

....

....

....

....

....

....

....

....

....

WHEEL-BALANCING

Unbalanced wheel and tyre assemblies adversely affect vehicle handling, stability and tyre wear.

These are two types of roadwheel imbalance:

.... and ....

A tyre and wheel assembly is in static balance when the mass of the assembly is uniformly distributed about its centre so that when mounted on a free bearing the assembly will come to rest in any position. If it always comes to rest in the same position, thus indicating a heavy spot, it is out of balance statically.

With the aid of a drawing describe the procedure for statically balancing a wheel assembly:

....

....

....

....

....

....

Dynamic balance

If the wheel shown statically balanced below is rotated at speed, certain out-of-balance forces will be produced, that is, the wheel is dynamically unbalanced.

Explain the reason for this and state the effect of the forces. Add arrows to the drawing to show the direction of the forces.

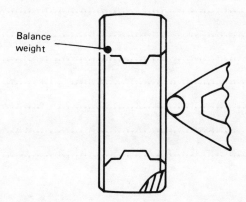

Balance weight

....

....

....

....

....

....

....

....

To dynamically balance the wheel shown below, additional weights are added to counteract the forces rocking the wheel. Show on the drawing where weights would be positioned to dynamically balance the wheel assembly.

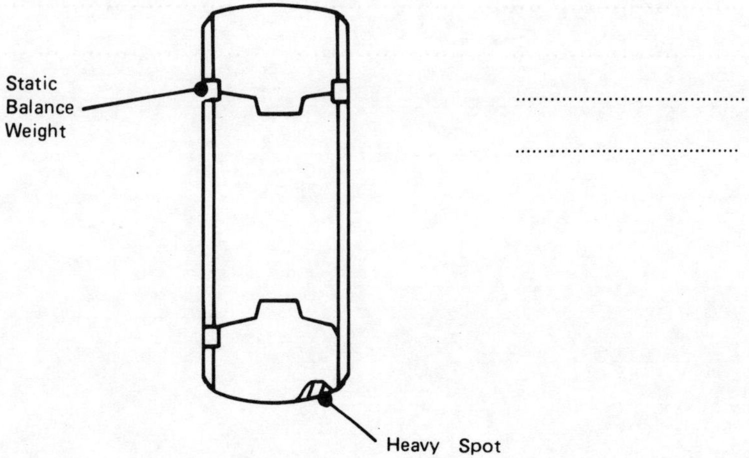

Static
Balance
Weight

....

....

Heavy   Spot

Two systems of wheel-balancing are used:

1. ....

....

2. ....

....

One advantage of balancing the wheel on the vehicle is that the hub and brakedrum or disc are also taken into consideration.

Describe, by referring to equipment with which you are familiar, the procedure for dynamically balancing a wheel.

Wheel-balancing procedure

....

....

....

....

....

....

....

....

....

....

State the effects of roadwheel imbalance on vehicle handling:

....

....

....

....

TYRE PRESSURE

During driving, tyre pressure varies; this is due to the change in air .... within the tyre.

If the vehicle travels at high speed over a long distance the temperature of the air in the tyre will ....
This will cause the tyre pressure to ....

State a possible increase in tyre pressure after a long journey ....

GAS LAW CALCULATIONS

It is possible to calculate the actual change in pressure (or temperature) of the air in tyres by using the 'gas equation' from Boyle's and Charles's laws, this is:

Assuming the volume of air in a tyre remains constant, $V_1 = V_2$; therefore the only factors to consider are pressure and temperature, thus:

$$\frac{P_1}{T_1} = \frac{P_2}{T_2}$$

When using the gas equation the pressure and temperature are 'absolute' values, that is

pressure = gauge pressure + ...

temperature K = °C + ...

Example

1. At the start of a journey the gauge pressure in a tyre is 196 kN/m² at a temperature of 15°C. Calculate the increase in tyre pressure at the end of the journey if the temperature of the air in the tyre had increased to 35°C.

2. In cold conditions the tyres of a vehicle are at a temperature of 7°C and a gauge pressure of 2 bar. What will be their pressure reading if after standing in the hot sun their temperature rises to 28°C?

Tyre pressures may be set correctly, but can rise substantially during a journey. What causes this increase in pressure and what should be done about it?

....

....

....

Chapter 9

Braking

ELEMENTS 24/43/44/45 **UNITS 17/37/38**

BRAKES

The purpose of a braking system on a vehicle is to:

1. Stop the vehicle.

2. ....

3. ....

FRICTION is use   to reduce the speed of a vehicle and bring it to rest. Friction material forced into rubbing contact with a rotating drum or disc, forming part of the wheel/hub assembly, generates HEAT. The energy of motion, that is, kinetic energy, is therefore converted into heat energy during the braking process. The heat energy is dissipated into the atmosphere. In addition to frictional contact at the braking surfaces, the ability to retard or stop a vehicle depends also on frictional contact at the ....
....

One of the functional requirements of a braking system is that the applying force, with which the brake surfaces are pressed together, should not require excess effort by the driver.

By what means is the driver's effort multiplied and transmitted to the braking surfaces?

....
....
....

....

System layout (hydraulic)

The drawing top right shows the master cylinder and front and rear drum brake assemblies for a light vehicle. Complete the drawing to show the piping layout for the system.

Hydraulic braking system

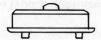

When the brake pedal is depressed, hydraulic fluid is displaced from the master cylinder into the system. The resulting increase of fluid in the wheel cylinders moves pistons which actuate the brake shoes and force the friction linings into contact with the rotating drum.

The parking brake on most cars operates on two wheels only and is normally mechanically actuated by cables. Examine a vehicle and sketch the handbrake cable layout below (show any cable guides).

DRUM BRAKES

(a)

(b)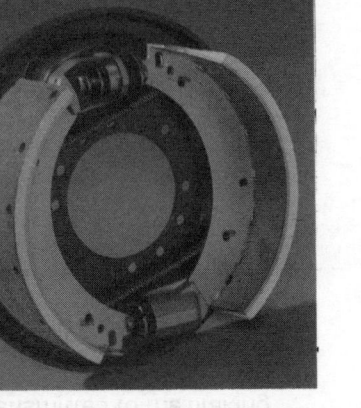

....

Label and name the drum brake assemblies shown above.
Briefly describe the application and action of each.

(a) ....
....
....
....
....
....

(b) ....
....
....
....
....
....
....

Name the brake lining/shoe fitments shown below.

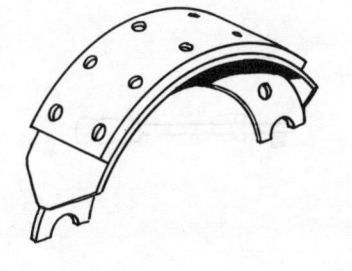 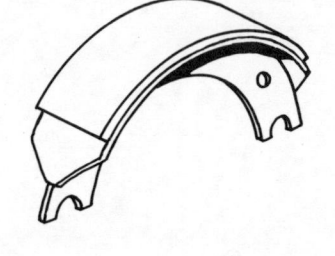

....

Friction materials

Materials used for friction linings must meet certain operational requirements, for example:

1. Good resistance to wear.

2. ....
....

3. ....

Resin-bonded asbestos has been the basic friction material used in braking systems; however, 'asbestos-free' materials are now superseding asbestos.

List some asbestos-free friction materials currently used in vehicle braking systems:

....
....
....

State a typical coefficient of friction for a brake friction material:

....

The drawing below shows a brake assembly incorporating a strut and bell-crank-lever mechanism. State the purpose of this mechanism, complete the labelling on the drawing and describe the operation of this brake.

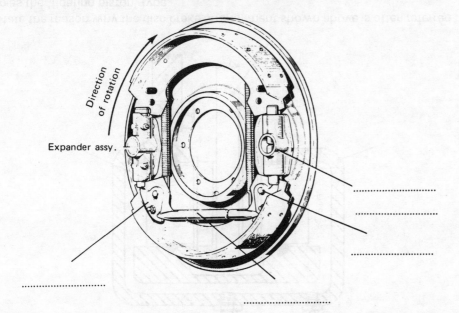

Direction of rotation

Expander assy.

....

....

....

....

....

The 'Duo-Servo' brake shown below incorporates a double-piston wheel cylinder and a floating adjuster mehchanism.

State the reason why this type of brake is referred to as 'Duo-Servo' and describe the action of the brake.

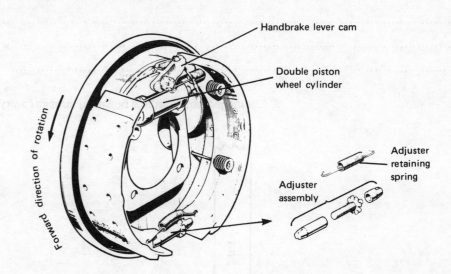

Handbrake lever cam

Double piston wheel cylinder

Adjuster retaining spring

Adjuster assembly

Forward direction of rotation

....
....
....
....
....
....

....
....
....
....
....
....

DISC BRAKES

The problem of brake fade can be overcome by the use of disc brakes (the principle of which is explained in the Level 1 book, page 49). Briefly, the advantages are greater heat dissipation, light weight, and discs tend to be self-cleaning. The disc which is bolted to the wheel hub is made from cast iron or cast steel while the calipher assembly is bolted to the stub axle assembly and straddles the disc.

Fixed-caliper (floating-piston) type

In the disc brake assembly shown below, a common fluid pressure applied to the two pistons forces the friction pads against the disc. Label the drawing.

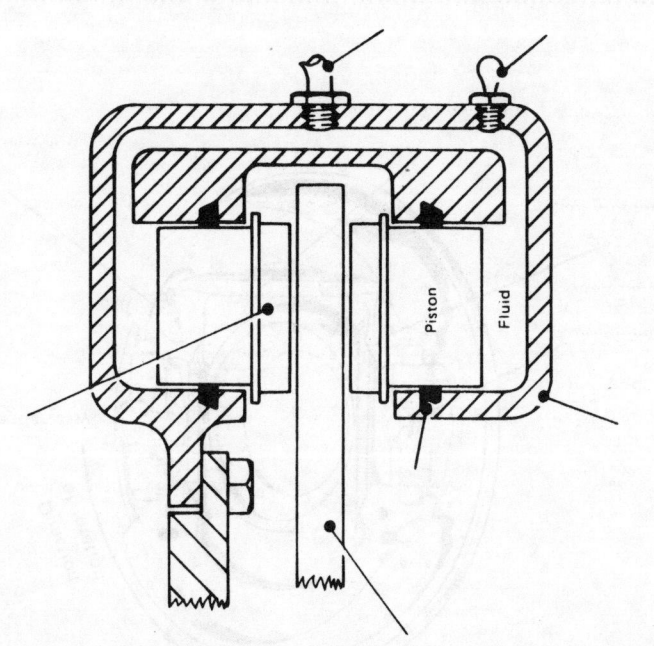

Piston

Fluid

State the reason why the disc brake arrangement shown above is often referred to as the 'floating piston' type.

....

....

Sliding or swinging-caliper type

In this design of disc brake the cylinder is free to slide or swing on a housing attached to the stub axle. When the brake is applied the increase in fluid pressure forces the piston in one direction and the cylinder in the opposite direction. Complete the drawing below to show one such arrangement and describe briefly how it operates.

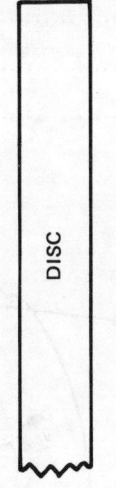

DISC

Give reasons for using single-piston calipers:

....

....

....

....

State briefly what is meant by 'brake fade':

....

....

PARKING BRAKE

Examine a handbrake lever, pawl and ratchet assembly and make a sketch in the space below to show the arrangement.

Compensating mechanism

A compensator is a 'balancing' mechanism in the parking brake system which ensures that each brake unit receives the same applying force. Examine a car or hgv and make a sketch of the parking brake compensating mechanism.

The sketch below illustrates a simple 'scissor' type parking brake actuator as used on many cars. Describe briefly the action of this mechanism.

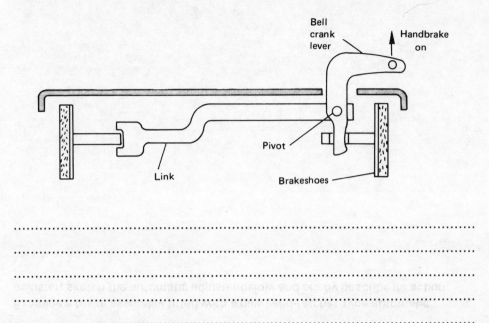

....

....

....

....

....

A hgv parking brake actuator is partly illustrated below; examine this arrangement and complete the drawing.

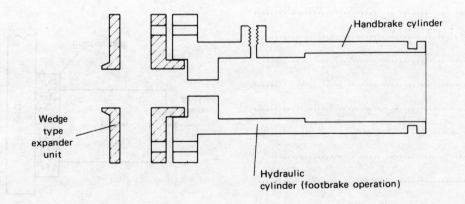

173

Automatic brake adjusters

When linings wear, some form of adjustment is necessary to bring the shoes nearer to the drum, otherwise excessive pedal travel will occur.

To eliminate the need for manual adjustment many vehicles are now equipped with automatic adjusters. These usually operate on the pawl-and-ratchet principle. However, they can suffer from the disadvantages of becoming rusted or seized due to inactivity.

A simple friction-type automatic adjuster is shown below; describe how this maintains the correct brake-shoe-to-drum clearance.

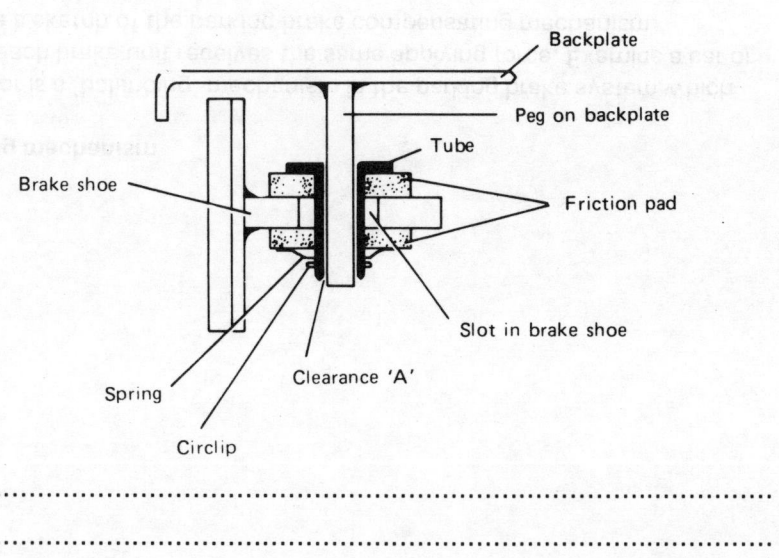

....

....

....

Disc brake adjuster

In most disc brake arrangements it is the rubber seal around the piston that maintains the correct clearance between pad and disc; it therefore serves two purposes, that is, it is a fluid seal and an automatic brake adjuster. Complete the simple sketch at the top opposite and describe the action of the rubber seal in a disc brake caliper assembly, in particular how it acts as an automatic adjuster.

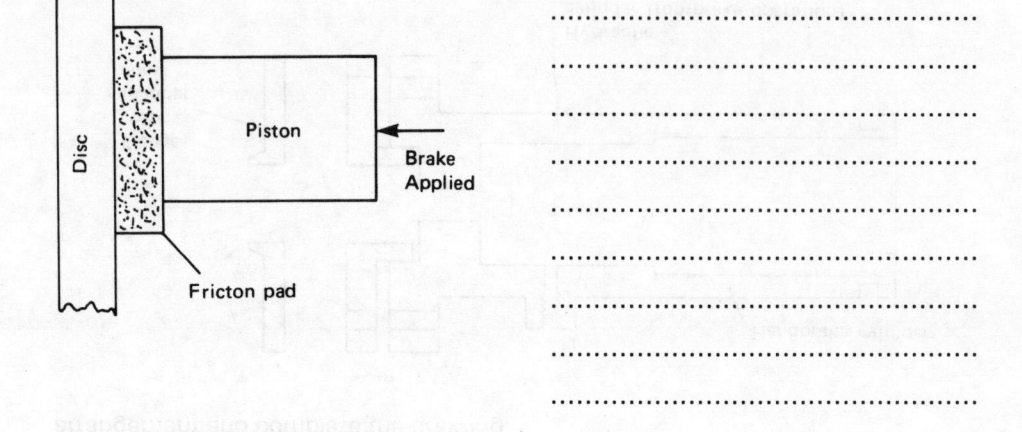

....

....

....

....

....

....

....

....

INVESTIGATION

Examine a brake assembly fitted with a pawl-and-ratchet-type automatic adjuster, sketch the automatic adjuster below and briefly describe its action.

HYDRAULIC BRAKES – MASTER CYLINDER

A master cylinder of the type shown below is usually made from cast aluminium alloy. It is mounted on the engine side of the bulkhead and operated by a pendant pedal.

Pedal depressed

The piston is pushed along the cylinder by the push rod. Once the piston main cup has covered the by-pass port the fluid is under pressure. When the pressure exceeds the residual pressure in the pipe lines, the rubber cup in the check valve opens and pressure is applied throughout the system.

Pedal released ....

....

....

....

INVESTIGATION

Examine a check valve and sketch two views, one when the rubber cup is lifted to allow fluid to flow out into the system and a second showing the check valve allowing the return of fluid.

The casing of a further type of master cylinder is shown below. This is a centre valve type. Carefully inspect such a cylinder then complete the sketch adding labels to the main components.

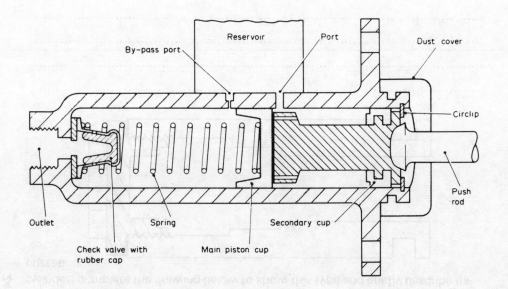

Reservoir — Port — Dust cover — By-pass port — Circlip — Push rod — Secondary cup — Outlet — Spring — Main piston cup — Check valve with rubber cap

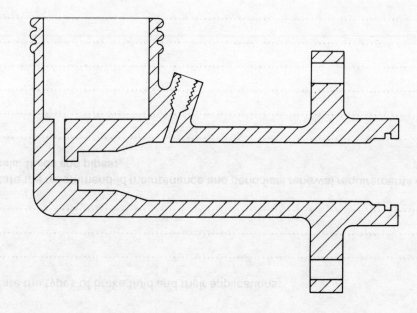

Many commercial vehicles employ the 'compression barrel'-type master cylinder; complete the drawing below to show this type and briefly describe its operation.

....
....
....
....
....
....
....
....
....
....

Brake fluid

This is in most cases a vegetable oil which, together with certain additives, gives the advantages of a low freezing-point, a high boiling-point and maintains a constant viscosity across a wide temperature range. Brake fluid must also be compatible both with metallic parts and with rubber seals used in the system. State the precautions necessary with hygroscopic brake fluid.

....

State the types of brake fluid and their applications:

....
....
....

State the recommended maintenance and periodical renewal requirements of seals, fluids and pipes:

....
....
....
....
....
....
....
....

Wheel cylinders

The wheel cylinder consists basically of a plain cylinder and piston which, energised by fluid pressure, actuates the brake shoe. A double-piston type is shown below. The diagram on the right is a cylinder for a two leading-shoe layout. Complete the diagram and label the important parts.

Rubber cup

Fluid entry

Rubber dust cover

Bleed hole

Types of hydraulic braking system

Hydraulic braking systems are either single-line, split-line or dual-line. Split- and dual-line systems necessitate the use of a .... master cylinder. In a split-line system the system is divided into two independent circuits, one to the front brakes and one to the rear brakes. Alternatively the 'split' can be diagonal, one front and one rear, or even triangular. Thus at least 50% of the wheels can be braked in the event of a failure in any part of the system.

A typical tandem master cylinder is shown below; complete the diagram; label the essential parts.

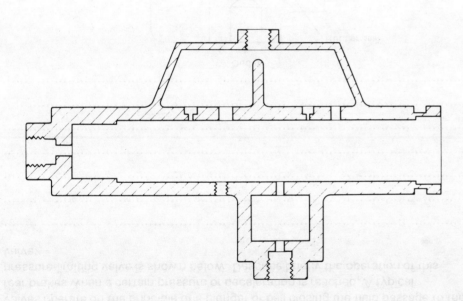

Complete the drawing opposite to show split- and dual-line braking system layouts.

The tandem master cylinder is so constructed that the front and rear brake chambers each have their own reservoir. When the push rod and main piston move along the cylinder the piston cup seals the by-pass port and places the fluid under pressure. Thus the secondary piston is moved and the fluid in this chamber is under pressure. At a sufficient pressure both check valves open and the wheel cylinder plungers are operated. Should a collapse of pressure occur in the rear brake chamber the secondary piston will travel until it meets the stop in its cylinder. Further brake pedal movement will build up pressure in the front chamber.

Split-line system

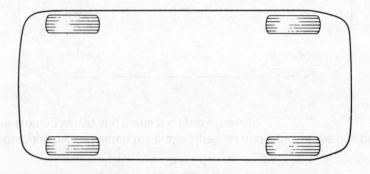

Dual-line system

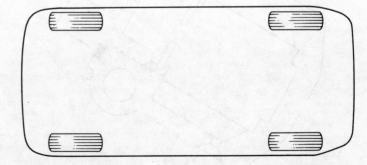

In most dual-line systems, drum brakes are employed at the rear and four piston disc brake calipers at the front.

PRESSURE-LIMITING VALVES

On modern cars, particularly those with front-engine and front-wheel drive, it is possible for the rear wheels to lock when under heavy braking. To overcome this problem a pressure-limiting valve can be fitted into the braking system. These valves operate on the principle of a plunger or ball closing the fluid passage to the rear brakes when a certain pressure or deceleration is reached. A typical pressure-limiting valve is shown below. Describe briefly the operation of this valve:

....

....

....

....

....

....

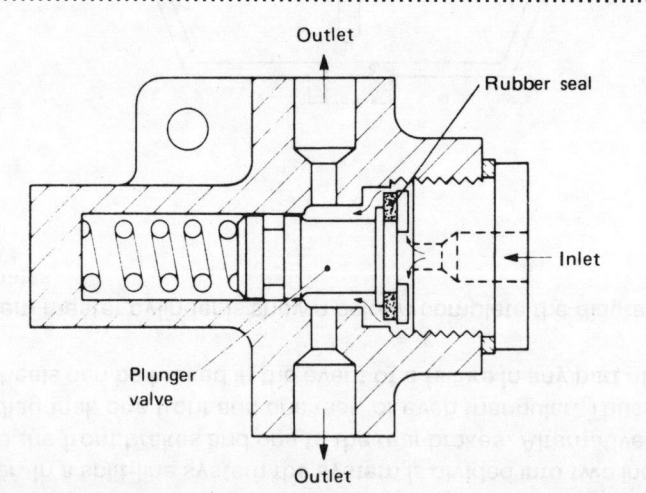

A second type of pressure-reducing valve is one that employs a cylinder which is set at a pre-determined angle. Within the cylinder is a steel ball which rolls up the incline, and at certain rate of deceleration closes the fluid outlet to the rear brakes. Complete the drawing top opposite to show this type of valve.

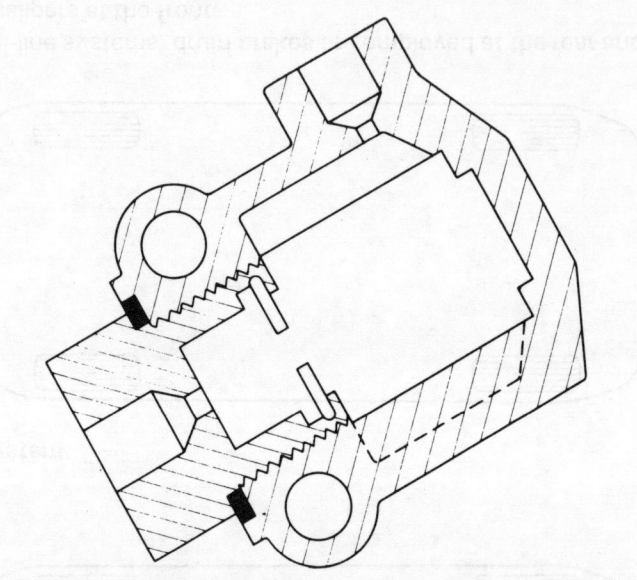

On the drawing below sketch the brake pipelines and indicate the position of the pressure-limiting valve, and name the type of valve.

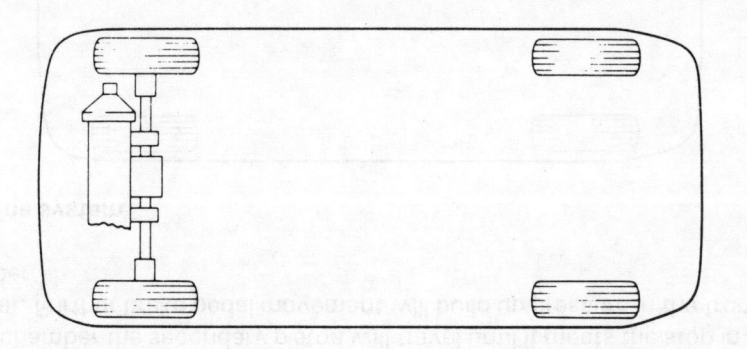

Give one advantage of the ball-type pressure-limiting valve over the plunger-type:

....

....

BRAKE VACUUM SERVO UNITS

The function of a brake servo unit is to augment the driver's pedal effort, thereby keeping the leverage required (and hence pedal travel) to a minimum.

General principle

The necessary pedal assistance is obtained by creating a 'pressure difference' across a large-diameter diaphragm or piston; the force of which, when applied to a hydraulic piston, increases the brake line pressure.

To obtain an air pressure difference a vacuum is created within the servo.

How is this achieved on a spark-ignition engine?

....

Describe the operation of the brake servo shown opposite:

....
....
....
....
....
....
....
....
....

Why is the servo known as a 'suspended vacuum' type?

....
....
....

Direct-acting brake servo

The drawing below shows a direct-acting suspended vacuum servo working in conjunction with a tandem master cylinder.

Complete the lower drawing to show the position of the control valve when the brakes are fully applied.

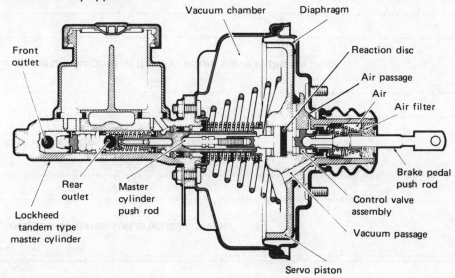

Brakes fully applied

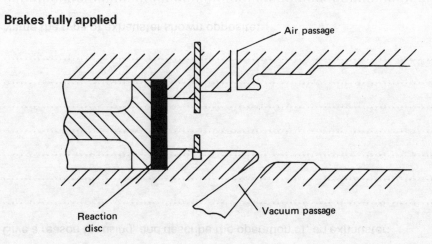

179

Many commercial vehicles employ larger-type suspended vacuum servo units which are activated from the brake pedal via a hydraulic master cylinder or by mechanical means; also included in certain systems are an engine-driven 'exhauster' and a 'vacuum reservoir'.

Complete the drawing below of a vacuum-assisted hydraulic braking system by adding the pipes and labelling.

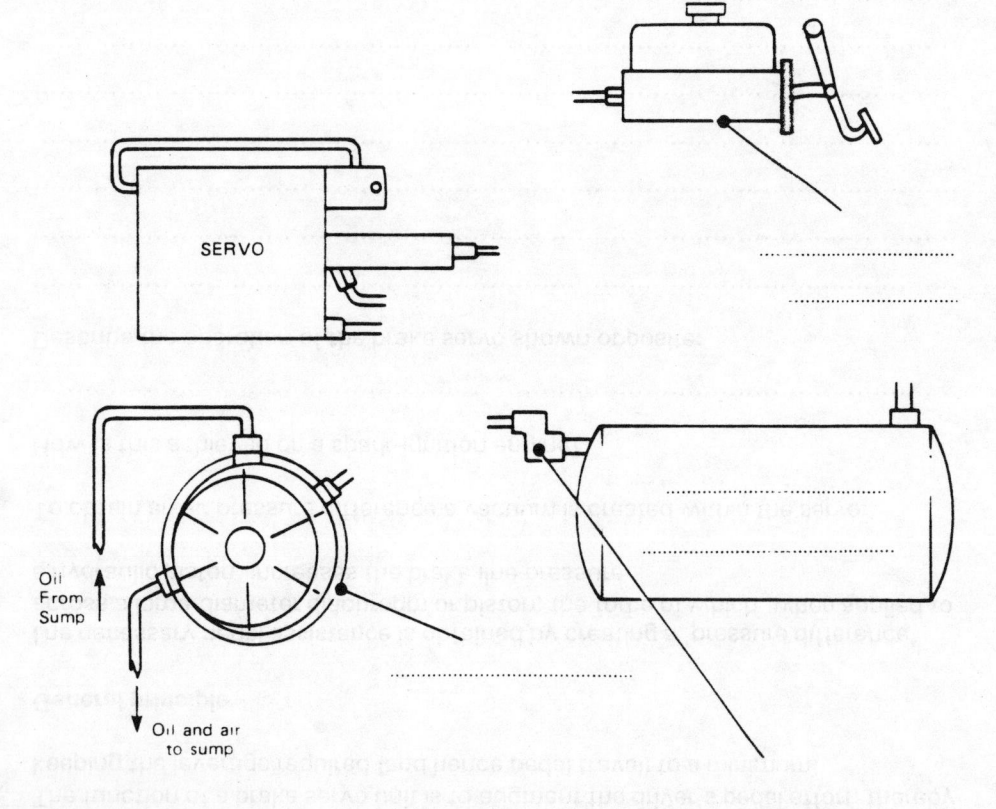

SERVO

Oil From Sump

Oil and air to sump

Give a reason for using, and describe the operation of, an exhauster:

....

....

....

....

....

....

....

Name the type of exhauster shown opposite:

....

....

State how an exhauster is usually driven:

....

....

State the purpose of the vacuum reservoir:

....

....

Describe the purpose of the non-return valve in the system:

....

....

In the arrangement shown opposite, oil is drawn into the exhauster unit. State the reason for this:

....

....

INDICATING DEVICES

Examine a vehicle and complete the drawings below to show the indicating devices and circuits.

STOP LAMP ACTUATION

PAD/BRAKE SHOE WEAR INDICATION

FLUID LEVEL

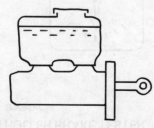

BRAKE TESTING

A brake test is carried out on a vehicle to determine the 'brake efficiency'. The efficiency is an indication of the ability of the braking system to stop the vehicle.

The brake efficiency of a vehicle is usually determined by:

(1) Measuring the rate of deceleration of a vehicle, or

(2) ....

Method (1) above involves the use of a ....

Method (2) above involves the use of a ....

List the main statutory requirements for braking systems relating to minimum efficiencies and MOT test requirements:

....
....
....
....
....
....
....
....
....
....
....

Describe the DECELEROMETER method and the ROLLER BRAKE TESTER method of determining the brake efficiencies for a vehicle.

DECELEROMETER METHOD (TAPLEYMETER)

....
....
....
....
....
....
....
....
....

ROLLER BRAKE TESTER

....
....
....
....
....
....
....
....
....
....
....
....

The somewhat obsolete decelerometer is still used in brake efficiency testing of certain vehicles. For which vehicles is this form of testing used and why?

....
....
....
....
....
....

In addition to normal brake efficiency tests, vehicles with ABS require rather more sophisticated test procedures to ensure the operation of the anti-lock braking system.
This calls for the use of dedicated or specific test equipment. Describe such equipment and testing carried out on anti-lock braking systems.

....
....
....
....
....
....
....
....
....
....

HYDRAULIC POWER BRAKING SYSTEMS

Hydraulic power, rather than compressed air or vacuum assistance, can be utilised to provide an applying force rather greater than that available by normal pedal effort.

Hydraulic booster system

This system entails the use of:

Fluid reservoir
Hydraulic pump (mechanically or electrically driven)
Accumulator
Valve block

Complete the labelling and describe the operation of the simplified layout below.

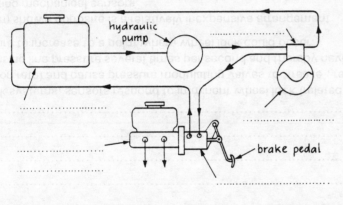

....
....
....
....
....
....
....
....

Alternatively the fluid can flow directly from the accumulator to the braking system (dynamic flow) via a driver's foot valve.

Describe the operation of the hgv 'full power' hydraulic system below; include in the description the construction and action of the hydraulic accumulator.

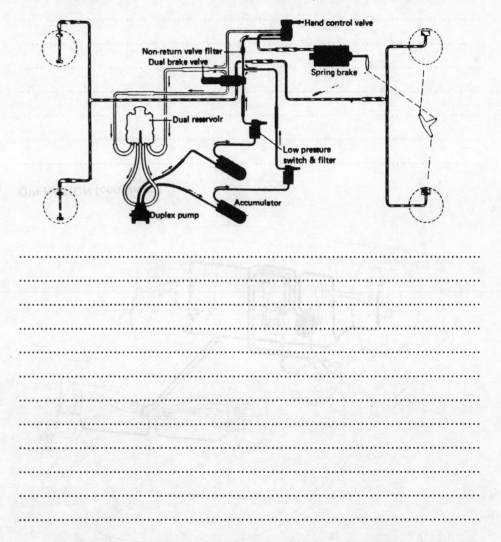

....
....
....
....
....
....
....
....
....
....
....

183

ANTI-LOCK BRAKING SYSTEM

To achieve maximum retardation and maintain steering control during braking, the road wheels must be on the point of locking or skidding. If skidding occurs, adhesion is reduced and the steering becomes ineffective, that is, control is lost. Anti-lock braking systems control the braking torque at the road wheels to a level which is acceptable according to the tyre/road adhesion, that is, the braking torque applied is the maximum permissible without locking-up the wheels. The main components in an anti-lock system are:

Hydraulic pump (mechanically or electrically driven).

....

....

....

....

In anti-lock systems, sensors respond to imminent wheel lock (related to deceleration rate) and cause pressure modulating valves to 'freeze', reducing and increasing line pressure several times per second and thereby never allowing the pressure to increase to a point where wheel lock could occur.

The system shown opposite is a relatively inexpensive arrangement incorporating mechanical sensors.

Label the drawing and describe the operation of the system.

OPERATION

....

....

....

....

....

....

....

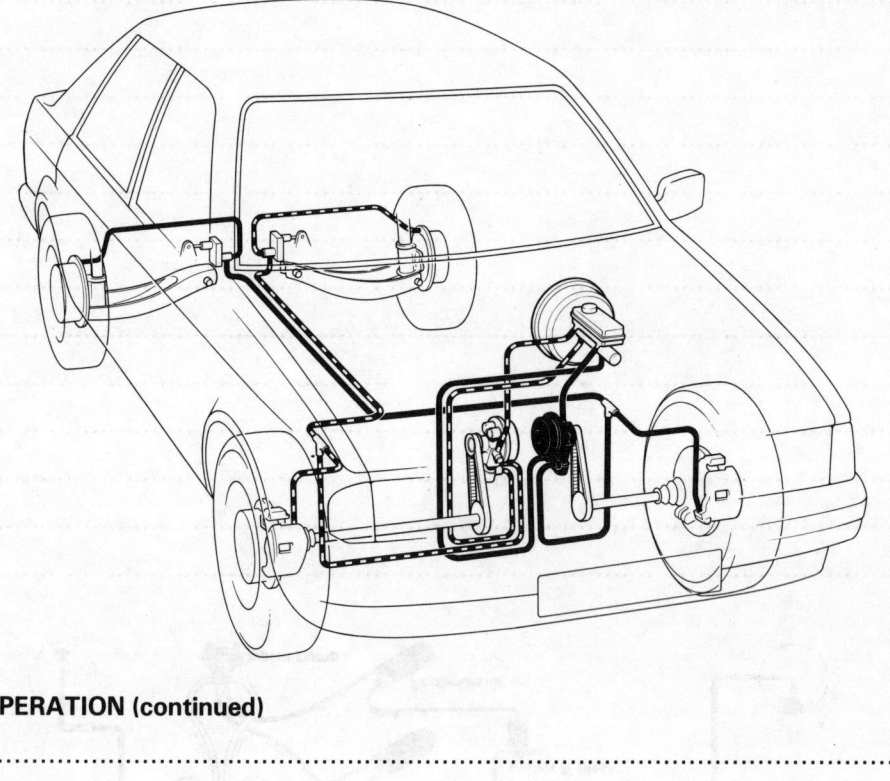

OPERATION (continued)

....

....

....

....

....

....

....

....

The ABS system shown opposite incorporates a power hydraulic booster. The valve block contains an input and an output valve for the rear brakes, separate input and output valves for each front brake, that is, the front brakes have independent anti-lock control, and a master valve.

Operation (normal braking)

The master cylinder actuates (via the driver's pedal) the front brakes and simultaneous operation of a spool valve puts the accumulator in direct communication with the rear brakes and master cylinder booster.
Describe the operation of the system during ABS regulation:

....
....
....
....
....
....
....
....
....
....
....
....
....
....

1. ....   2. ....
3. ....   4. ....
5. ....   6. ....
7. ....   8. ....

Complete the labelling on the drawing.

A _____   ....
B ------   ....
C ·········   ....

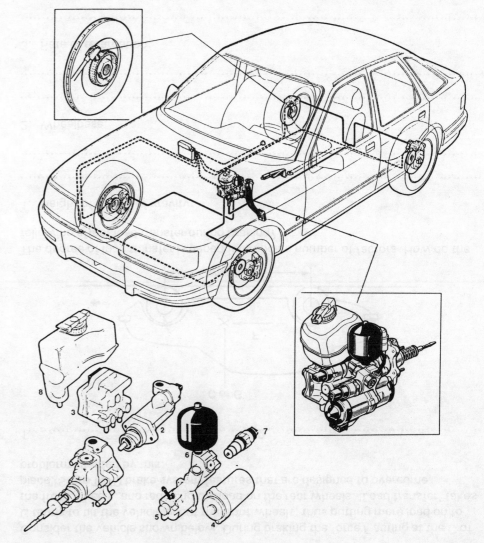

24.3 CENTRE OF GRAVITY AND LOAD TRANSFER

On the vehicle shown below the cross marks the *centre of gravity*. With a front engine, front-wheel-drive vehicle the C of G would be nearer to the front. However, with a rear engine, rear-wheel-drive vehicle the C of G would be nearer to the rear of the vehicle.

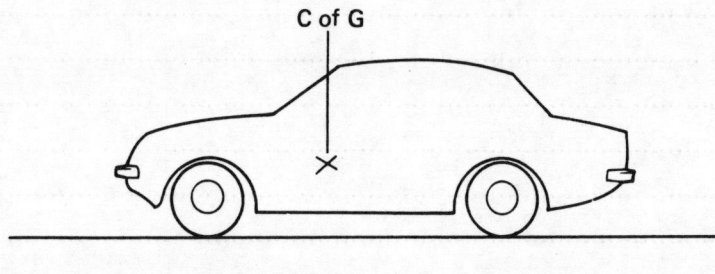

C of G

Briefly describe what is meant by the term 'centre of gravity':

....
....
....
....
....

The stability of a vehicle when braking (or cornering) is greatly affected by the actual position of the vehicles C of G.

Give examples of vehicle types which have:

1. Low C of G ....
2. High C of G ....

What other vehicle factors can adversely affect braking stability?

....
....

Consider the vehicle shown below. During braking the force *F* acting at the C of G tends to tilt the vehicle about the front wheels, thus putting more load on to the front wheels, and reducing the load on the rear wheels. 'Load transfer' takes place. State TWO brake system features that are designed to overcome problems created by this:

1. .... 2. ....

C of G

F

Wheelbase

The degree of load transfer is dependent upon a number of factors. How do the following affect load transfer during braking?

1. Height of centre of gravity:

....
....

2. Wheelbase:

....
....

3. Rate of retardation:

....
....

4. Ratio of C of G height to wheelbase:

....
....

STOPPING DISTANCE

The stopping distance for a particular vehicle is dependent on a number of factors other than the design and condition of the vehicle's braking system.

List FIVE factors, external to the vehicle, which can affect stopping distance.

1. ....
2. ....
3. ....
4. ....
5. ....

Maximum retardation of a vehicle is achieved just before wheel lock (skidding) occurs. This is because, when sliding or skidding of the tyres on the road surface occurs, friction is reduced. This is known as *kinetic* or *sliding* friction.

Why is the braking force higher if the wheels continue to roll?

....
....
....
....
....

WORK DONE

When a vehicle is brought to a halt, the 'work done' by the braking system is a product of:

1. ....
2. ....

The formula used to calculate work done is:

The retarding force produced by the brakes when stopping a vehicle is 5000 N. If this force is applied over a distance of 25 m, calculate the work done by the brakes.

If the work done by the brakes in stopping a vehicle is 200 kJ, calculate the retarding force if the distance travelled during brake application is 40 m.

The work done by the brakes in stopping a vehicle is 600 kJ. If the retarding force produced by the brakes is 20 kN, the distance travelled by the vehicle during braking is:

(a) 3 m

(b) 30 m

(c) 60 m

(d) 90 m?

Answer (   )

187

Braking performance

Many factors affect the actual distance in which a vehicle can be brought to rest by the brakes.

State the effect on stopping distance of:

Vehicle speed

....

....

Vehicle mass

....

....

Road conditions

....

....

Velocity/Acceleration/Deceleration

Velocity is the rate at which a body moves in a given direction. The rate of movement is expressed in m/s^2.

The speed of a body is also the rate of movement; the difference between speed and velocity is that speed does not involve direction.

Define:

Acceleration

....

....

Deceleration

....

....

Acceleration due to gravity

The acceleration of a free-falling body due to the force of gravity is

....

In theory the maximum deceleration of a vehicle can only equal the acceleration due to gravity, since gravity maintains contact between tyres and road.

Braking efficiency

State the meaning of braking efficiency:

....

....

....

The braking efficiency of a vehicle can be determined by:

(a) expressing the deceleration as a percentage of $9.81 \ m/s^2$
or

(b) expressing the braking force as a percentage of the

....

If a vehicle decelerates at $9.81 \ m/s^2$, the braking efficiency is said to be ....%.

For the braking force to be equal to the vehicle weight the coefficient of friction between the tyre and the road would have to be ....

Using method (a) above:

Braking efficiency $= \qquad \times \dfrac{100}{1}$

Using method (b) above:

Braking efficiency $= \qquad \times \dfrac{100}{1}$

During a brake test the maximum rate of deceleration for a vehicle was 7.5 m/s^2. Calculate the braking efficiency.

When testing the parking brake on the rear axle of a vehicle, the efficiency was found to be 27%.

Calculate the load on the rear wheels if the recorded braking force was 4 kN.

A vehicle has a mass of 2000 kg. Calculate the braking efficiency if the total retarding force during brake application is 14 kN.

Calculate the rate of deceleration of a vehicle during braking if the total retarding force is 11.5 kN and the vehicle mass is 1600 kg.

A vehicle is uniformly decelerated at 4 m/s^2. Calculate the retarding force and the braking efficiency if the vehicle has a mass of 1000 kg.

Complete the table:

% Efficiency	Vehicle mass (kg)	Braking force (kN)	Decel. (m/s^2)
25		5	
	1500	12	
	2300		6.5
72	3000		

189

<h1>24.3 VEHICLE SPEED AND STOPPING DISTANCE</h1>

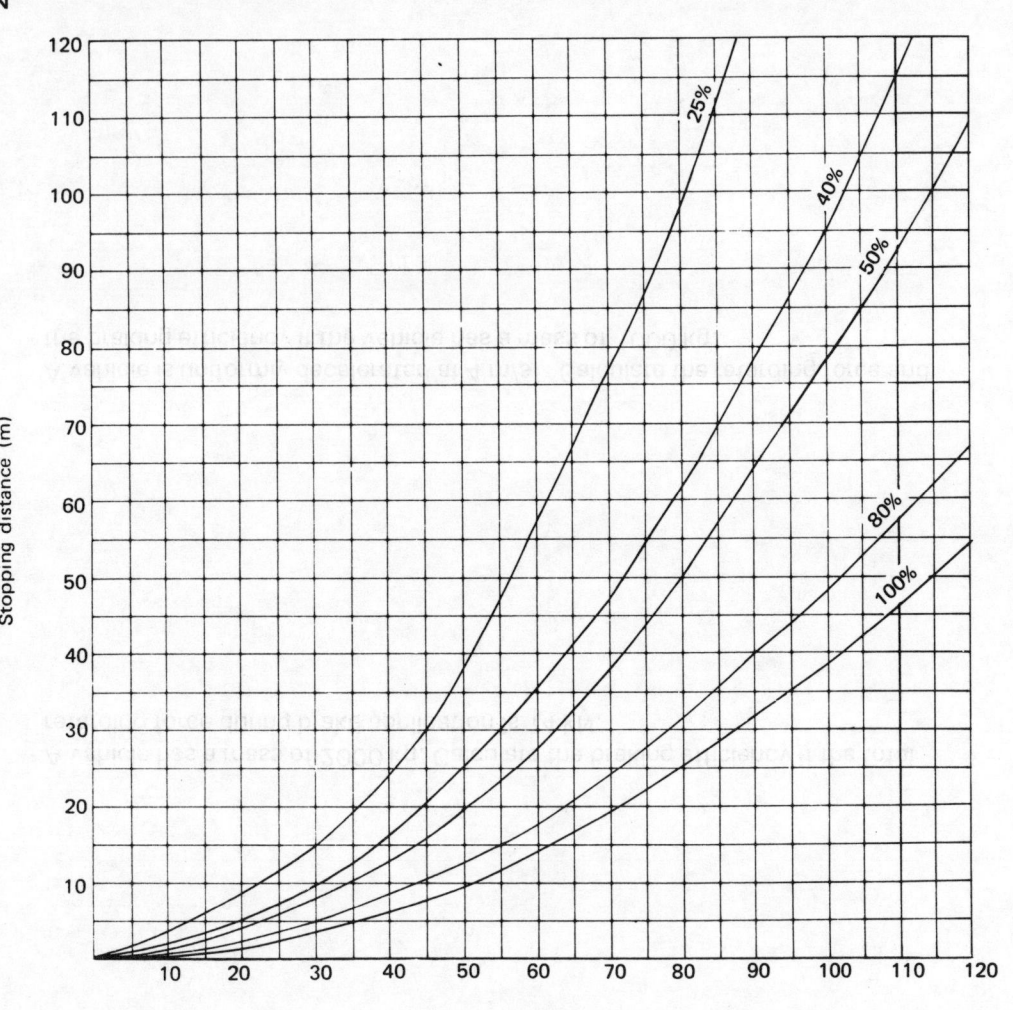

Stopping distance (m) (vertical axis)

Vehicle speed (km/h) (horizontal axis)

It can be seen from the graph that, if the speed doubles, the stopping

distance is .... greater, that is, the stopping distance increases as

the .... of the speed.

The graph opposite illustrates the effect of vehicle speed on stopping distance.

Determine from the graph:

1. The braking efficiency if the stopping distance is 48 m from a speed of 100 km/h.

2. The stopping distance from a speed of 75 km/h if the braking efficiency is 40%.

From the figures given below, plot a curve on the graph opposite of stopping distance against vehicle speed and, from information already on the graph, state the brake efficiency for the vehicle (see note below).

Speed (km/h)	36	54	72	90	108
Stopping dist. (m)	7.7	17.3	30.8	48	69.25

It can be seen from the graph that the stopping distance is proportional to the efficiency, that is, if the braking efficiency is halved the stopping distance is doubled.

Therefore for a given speed, if the stopping distance with 100% efficiency is known, it is possible to establish the efficiency for a vehicle using this graph if its stopping distance is known, for example, from a graph.

Stopping distance at 95 km/h at 100% efficiency is 35 m.

Stopping distance at 95 km/h at 40% efficiency is 87.5 m.

Efficiency $= \dfrac{35}{87.5} \times \dfrac{100}{1} = 40\%$

When the brakes are applied on a vehicle, the hydraulic or air pressure acting on pistons or diaphragms forces the friction material into contact with the rotating drum or disc to produce a braking force and braking torque. It is important to appreciate the relationship between the pressure in the system, the area of the pistons or diaphragms and the force exerted by the pistons or diaphragms. State the relationship between PRESSURE, FORCE and AREA:

....

....

....

....

....

....

State the two factors that contribute to the ratio of movements and forces between the brake pedal and wheel cylinder pistons in a hydraulic braking system:

1. ....

2. ....

State the effect the inclusion of air in a hydraulic braking system would have on:

(a) the 'movement ratio'

....

....

(b) the 'force ratio'

....

....

INVESTIGATION

To show the relationship between movement ratio and force ratio in a simple hydraulic system.

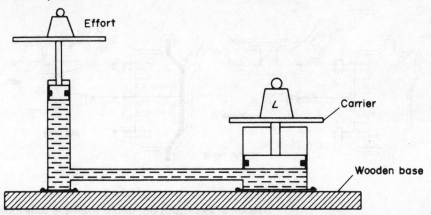

Use a simple hydraulic system similar to that shown above.

Force ratio

1. Place a load on the large piston carrier.

2. Determine by experimenting, the effort required to raise the load at a uniform speed.

$$\text{Load} =$$
$$\text{effort} =$$

$$\text{force ratio} = \frac{\text{load}}{\text{effort}} = \underline{\hspace{3cm}} = \dotsb$$

Movement ratio

Allow the load to rise a measured amount and determine the corresponding movement of the effort piston.

$$\text{Movement ratio} = \frac{\text{distance moved by effort}}{\text{distance moved by load}}$$

$$\text{movement ratio} = \underline{\hspace{3cm}} = \dotsb$$

Compressed air braking systems

In its simplest form a full air braking system consists of the following basic components: air compressor, reservoir, unloader valve, foot valve and brake chambers or actuators.

Outline the basic operation of the simplified single-line system shown opposite with particular reference to the function of the compressor, unloader valve and foot valve.

....
....
....
....
....
....
....
....
....

Sketch below an 'S' type cam with roller followers.

Single-line air brake system

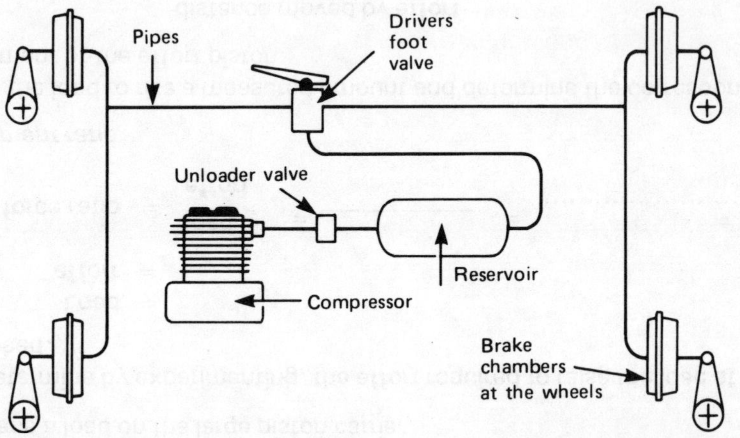

Complete the drawing on the right below to show the action of a single-diaphragm actuator when the footbrake is applied.

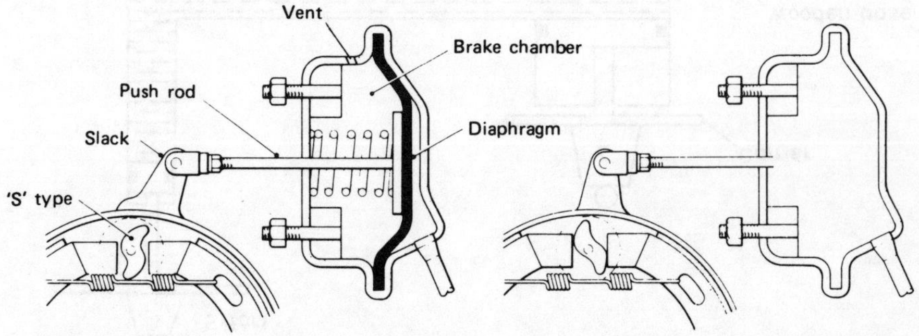

State the reason why the S-type cam is used in preference to the simple cam:

....
....

Split-line system

In the split system the front and rear brakes are independent systems operated from separate reservoirs via a dual foot valve. This system provides a secondary brake, operative on 50% of the wheels, in the event of a failure in one part of the system. Complete the labelling and add the pipes to the drawing below.

Dual brake valve

Res

Res

Single-check valves

Unloader and safety valve

State the purpose of the:

Single check valves

....

....

....

Safety valve

....

....

....

State the causes and effects of water accumulation in a compressed-air braking-system:

....

....

....

....

....

State the purpose and describe the action of an alcohol evaporator in an air braking system, and add one to the system opposite:

....

....

....

....

....

....

....

Some systems employ an ALCOHOL INJECTOR which automatically and frequently injects a quantity of alcohol into the system. Its function is to:

....

....

....

One method of overcoming the problem of water and other contaminants entering the main braking system is to employ an AIR DRIER. This unit is located between the compressor and reservoir, warm air entering the unit is cooled and the water content is absorbed into crystal-type material.

When the governor unloads the compressor, the water is automatically expelled from the air drier.

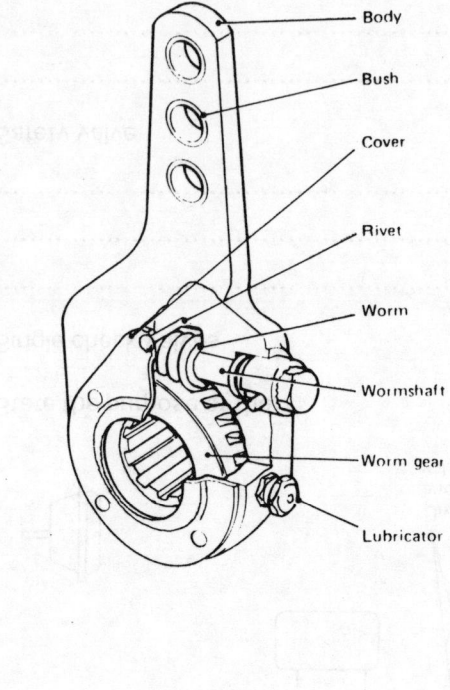

Body
Bush
Cover
Rivet
Worm
Wormshaft
Worm gear
Lubricator

The **SLACK ADJUSTER** lever shown left has a manual adjuster (worm and worm gear).

In the automatic version, the worm gear is on a ratchet and automatically takes up slack if the lever movement is excessive.
State the purpose of the component.

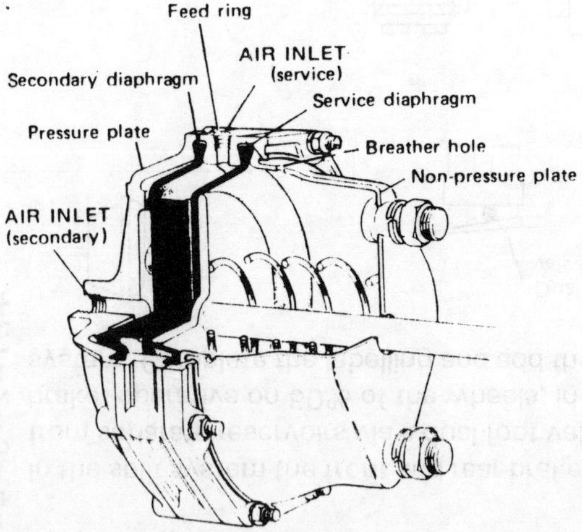

....
....
....
....
....
....
....

Feed ring
AIR INLET (service)
Secondary diaphragm
Service diaphragm
Pressure plate
Breather hole
Non-pressure plate
AIR INLET (secondary)

Name and describe the operation of the brake chamber shown on the left.

....
....
....
....
....
....
....
....

Dual-line system

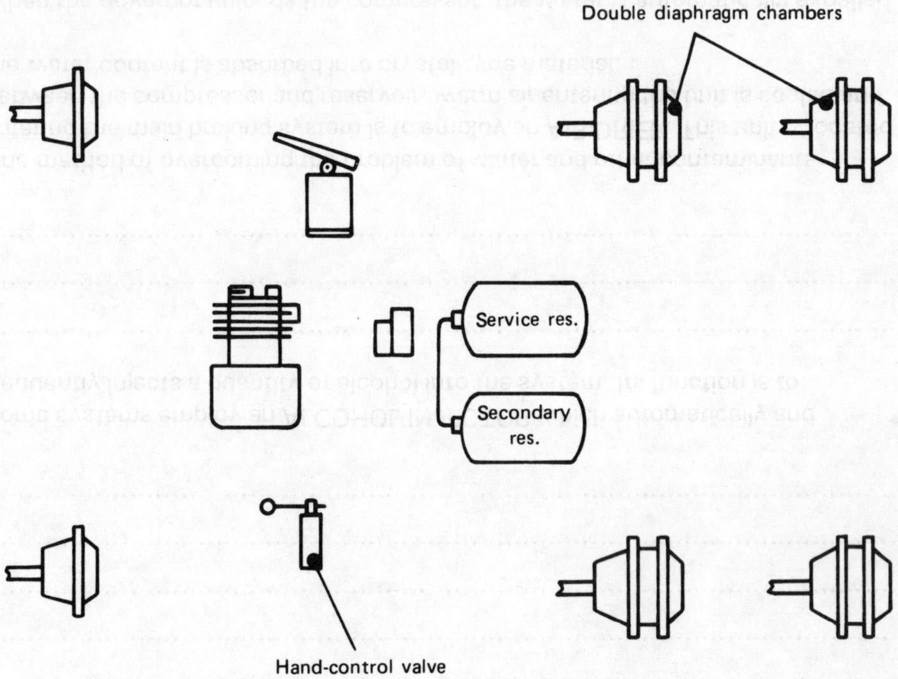

Double diaphragm chambers

Service res.

Secondary res.

Hand-control valve

A dual-line layout for a six-wheeled vehicle is shown above; add the pipes to the drawing and describe the operation of the system.

....
....
....
....
....
....
....
....

194

Quick release valve

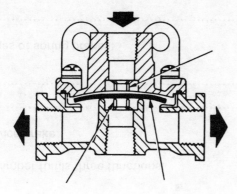

State the purpose and describe the operation of the valve shown above:

....

....

....

....

....

....

....

DRAIN VALVES in the system may be manual or automatic in operation. Why are these valves necessary?

....

....

....

Relay valves

Relay valves provide a means of admitting and releasing air to and from the brake chambers, in accordance with brake valve operation, without passing the air for the brake chambers through the brake valve.

This gives much quicker braking responses and also permits the use of smaller control valves in modern cab designs.

Add the pipes to the arrangement shown below to show how a relay valve would be incorporated into the system.

Pressure-regulating valve

The drawing below shows the location of a pressure-regulating valve in an air brake system; state the purpose of this valve.

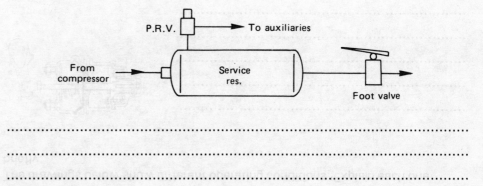

....

....

Spring-brake system

The spring-brake chamber is a single-diaphragm brake chamber with an extension cylinder on the rear containing a spring-loaded piston. Label the spring-brake chamber shown below.

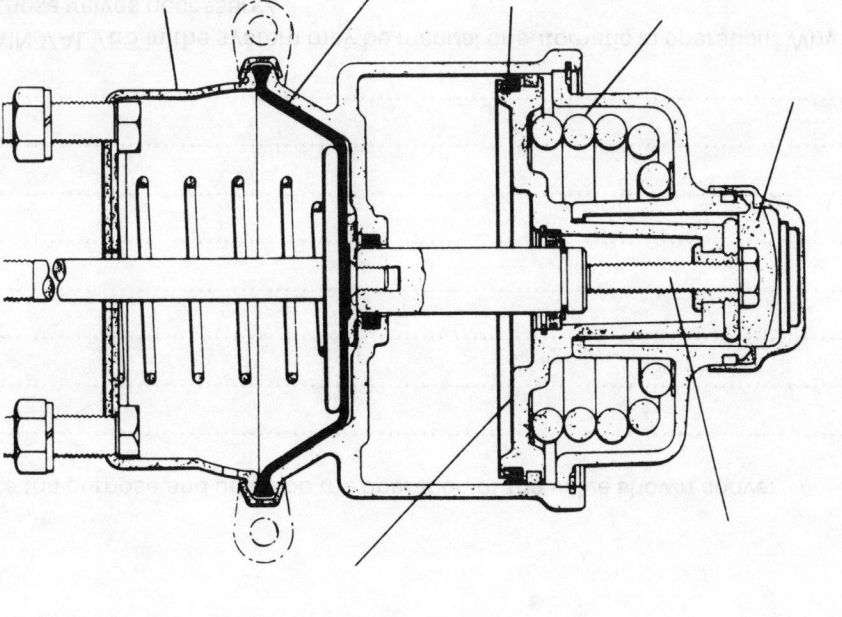

The spring-brake chamber fulfils three functions:

1. It provides a service brake.

2. ....

3. ....

Give three advantages of spring brakes:

1. ....

2. ....

3. ....

The drawings below show different operating conditions. Explain each one briefly.

Normal driving

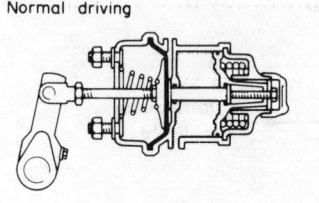

....
....
....
....

Service brake

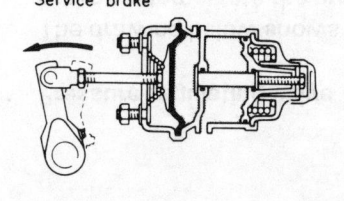

Service brake portion acts in similar manner to standard brake chamber. Air pressure supplied (normally) via foot valve. Spring is held compressed by a steady air pressure in spring chamber.

Secondary and parking

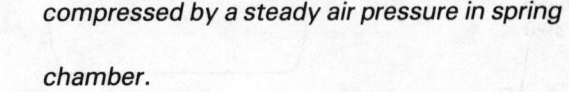

....
....
....
....

Manual release

....
....
....

THREE-LINE SYSTEM

The braking layout shown is a THREE-LINE system as used on many articulated vehicles and drawbar trailer combinations. Spring-brake chambers are employed on the tractor and double diaphragm units on the trailer.

Add the three-line trailer braking system to the drawing below and indicate the colour coding for the tractor/trailer connections.

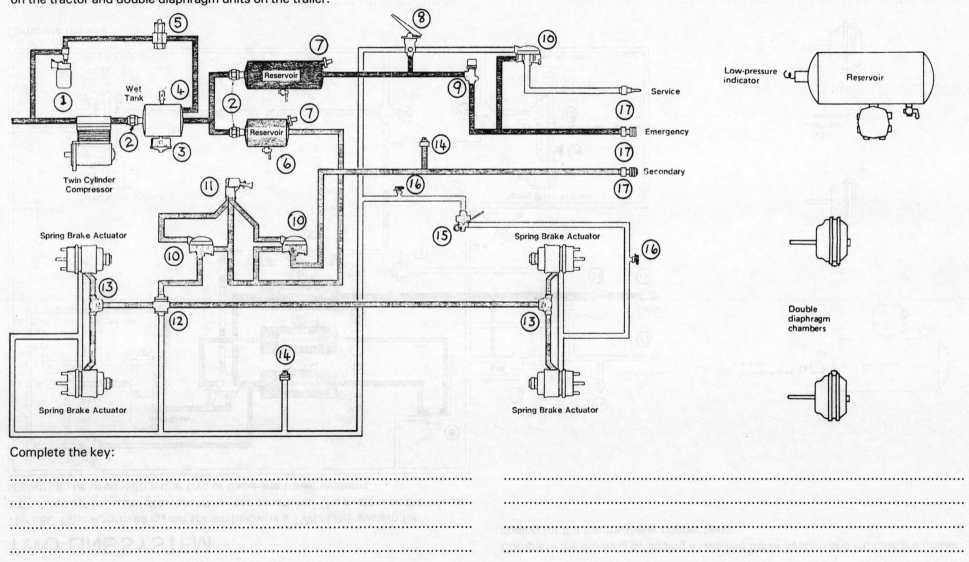

Twin Cylinder Compressor

Wet Tank

Reservoir

Reservoir

Spring Brake Actuator

Spring Brake Actuator

Spring Brake Actuator

Service

Emergency

Secondary

Low-pressure indicator

Reservoir

Double diaphragm chambers

Complete the key:

....

....

....

....

....

....

....

....

TWO-LINE SYSTEM

The EEC Type Approved layout shown below is a TWO-LINE system for articulated vehicles and drawbar trailer combinations. The system on the tractor shown can be used with either two or three-line trailer systems.

Complete the drawing by adding a two-line trailer system and indicate the colour coding for the tractor/trailer connections.

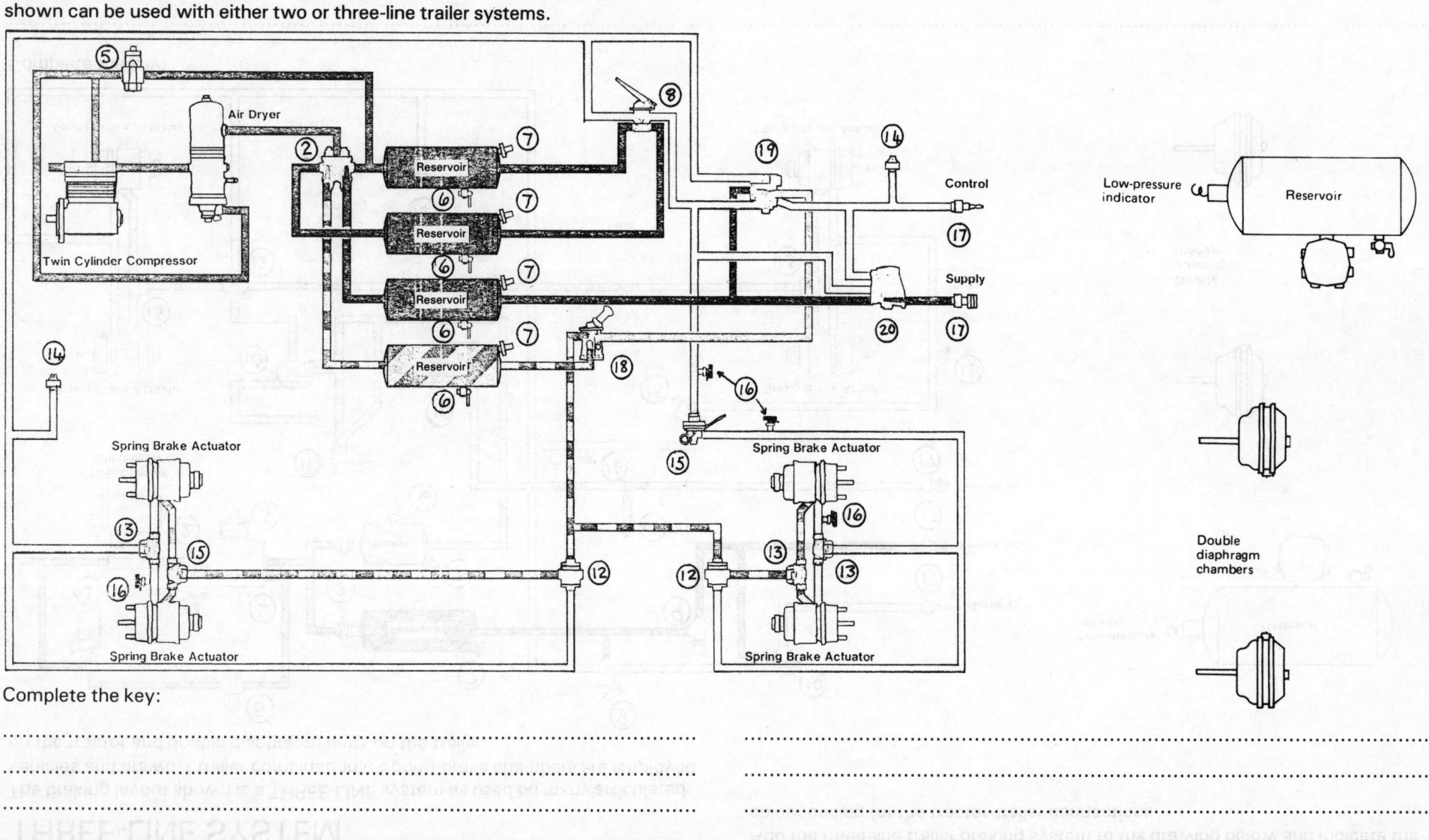

Air Dryer

Reservoir

Twin Cylinder Compressor

Control

Supply

Low-pressure indicator

Reservoir

Spring Brake Actuator

Spring Brake Actuator

Spring Brake Actuator

Spring Brake Actuator

Double diaphragm chambers

Complete the key:

....

....

....

Refer to system on page 197.

Describe the operation of the three-line system.

Service brake

....
....
....
....
....
....

Secondary/auxiliary brake

....
....
....
....
....

Parking brake

....
....
....
....
....

Emergency system

....
....
....
....

Refer to system on page 198.

Describe the operation of the two-line system.

Service brake

....
....
....
....
....
....

Secondary/auxiliary brake

....
....
....
....
....

Parking brake

....
....
....
....
....

Emergency system

....
....
....
....

SYSTEM COMPONENTS

State the purpose of the following:

Governor valve

....

....

....

Tractor protection valve

....

....

....

....

....

Differential protection valve

....

....

....

....

....

Describe the methods used to identify trailer air line connectors:

....

....

....

....

....

Double-check valve

This valve permits the air from two sources to flow down a common line when either system is operative.

When both systems are operated at the same time the double-check valve responds to the system registering the highest pressure.

The differential protection valve is a 'biased' type of double-check valve.

Complete the drawing below to show a simplified double-check valve.

Hand-valve supply → ← Foot-valve supply

To brake chambers

Make a simple diagram below to show where the valve above would be used in a brake system.

Lock actuators

The lock actuator is a double-diaphragm or diaphragm/piston brake chamber fulfilling the function of service and secondary braking. In addition the unit incorporates a device which enables the brakes to be 'mechanically' locked in the applied position when parking the vehicle, thus eliminating the conventional handbrake linkage.

Complete the labelling on the lock actuator shown below.

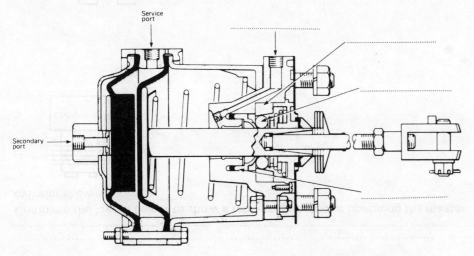

Service
port

Secondary
port

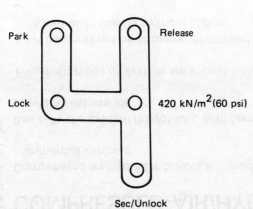

Park   Release

Lock   420 kN/m^2(60 psi)

Sec/Unlock

The gate for the full-flow secondary/park valve, which is used in the lock actuator system is shown left; describe, alongside the headings opposite, the operation of the lock actuator system.

Service brake

....
....
....
....

Secondary brake

....
....
....
....

Parking brake applied

....
....
....

Parking brake released

....
....
....

With the lever held in the unlock position, secondary line pressure is applied to the unloader or governor valve; this allows the secondary brake line pressure to build up beyond normal operating pressure (provided the engine is running) to free the locks under abnormal conditions.

COMPRESSED-AIR/HYDRAULIC BRAKES

Compressed-air/hydraulic brakes are used on many light/medium-range commercial vehicles.

Basically the system is hydraulic with some form of compressed-air unit providing assistance.

Two main types of system are in use:

1. A compressed-air actuator, controlled from a foot valve, operating directly on a hydraulic master cylinder piston.

2. ....

....

Complete the layout below to show a tandem air actuator operating the master cylinder shown.

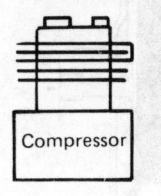

Compressor

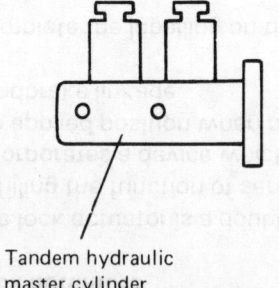

Tandem hydraulic
master cylinder

The layout below shows simplified a compressed-air servo unit operated mechanically from the footbrake.

Study the drawing and describe the operation of this system.

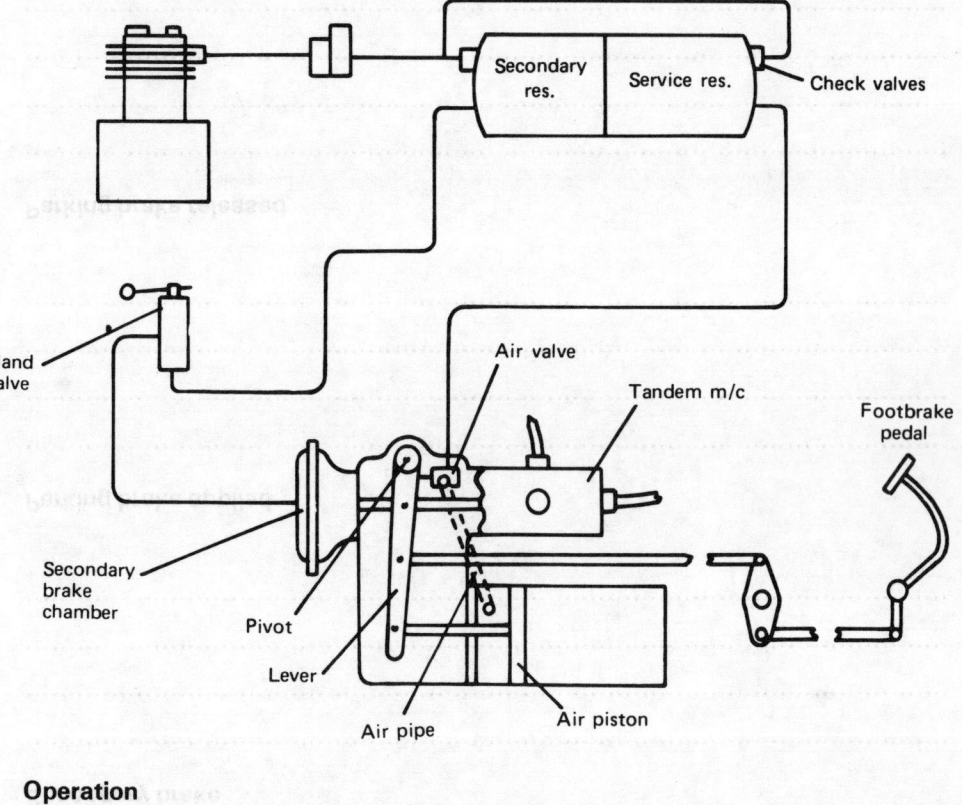

Operation

....

....

....

....

....

....

JACK-KNIFING

Jack-knifing is a problem associated with articulated vehicles. The drawing below illustrates what happens when jack-knifing occurs, that is, the tractor unit folds round on to the semi-trailer.

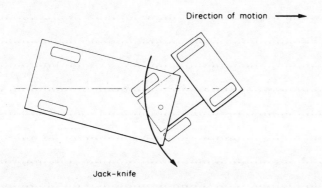

Direction of motion ➞

Jack-knife

When is jack-knifing most likely to occur and what are the main causes of the problem?

....
....
....
....

ANTI-JACK-KNIFE SYSTEM

1. A multi-disc brake can be incorporated into the fifth wheel coupling so that when the brakes are applied the fifth wheel coupling is effectively locked to the trailer king-pin. This is normally added as a modification to a standard vehicle.

2. ....

3. ....

ANTI-LOCK BRAKING

The system shown below is very popular on articulated tractor units; describe briefly how it prevents wheel lock during braking.

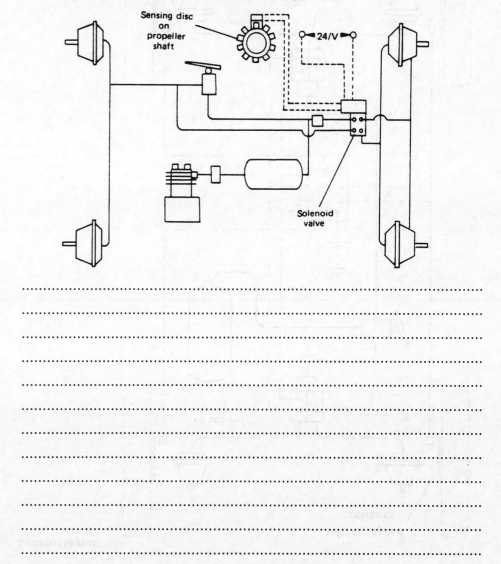

Sensing disc on propeller shaft

24/V

Solenoid valve

....
....
....
....
....
....
....
....
....
....
....

ABS (ANTILOCK-BRAKING SYSTEM) hgv

System layout

The drawing opposite shows the basic layout of a full air ABS system for a truck. This system, which is employed on rigid and articulated hgvs, prevents wheel lock during braking and automatically utilises the maximum road surface friction available. This shortens the stopping distance and maintains vehicle stability.

Describe the operation of the system shown.

....
....
....
....
....
....
....
....
....
....
....
....
....
....
....
....
....
....
....
....
....
....
....

wheel speed sensors

reservoir

ECU

pressure control valves

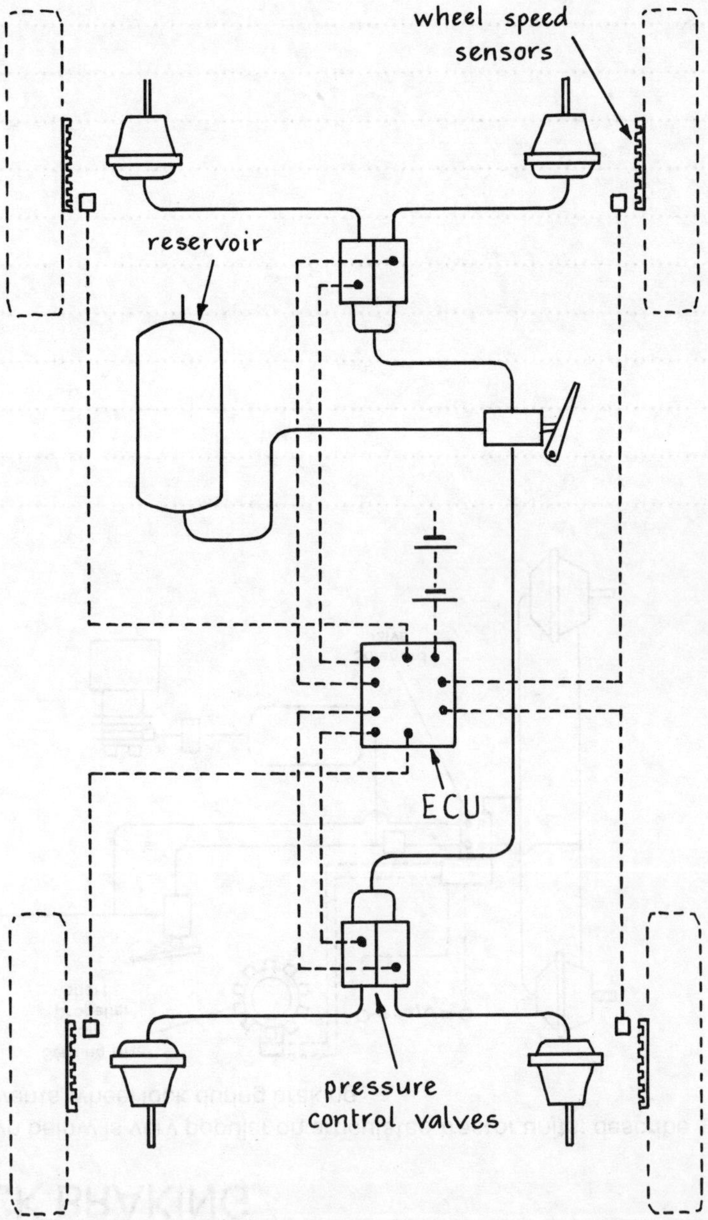

LOAD SENSING VALVES

Load sensing valves are used in the braking systems of cars and hgvs. The valves apportion the braking force at an axle according to the loading on the axle. It is necessary to reduce the braking force as the load on the axle reduces. Why is this?

....
....
....

State where load sensing valves would be used on cars and hgvs, and why.

....
....
....

Cars:

....
....

hgvs:

....
....
....

Add a load sensing valve to the drawing below.

Chassis

Operation

A schematic view of a load sensing valve is shown below; describe the operations of the valve.

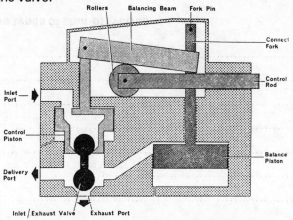

Rollers   Balancing Beam   Fork Pin

Connect Fork

Control Rod

Inlet Port

Control Piston

Balance Piston

Delivery Port

Inlet / Exhaust Valve   Exhaust Port

....
....
....
....
....
....
....
....
....

VEHICLE RETARDERS (AUXILIARY BRAKES)

A vehicle retarder is a device that slows down a vehicle without the driver having to operate the normal braking system.

Retarders were originally used on heavy vehicles and coaches operating on journeys which involved long descents (for example, mountainous roads on the Continent). Using a retarder a driver can control the speed of a vehicle during a long descent without the need for continuous brake application which can cause brake fade.

However, owing to improvements in retarders, increased vehicle speeds and motorway operation, many more advantages are to be gained by using vehicle retarders and their fitment has become widespread.

List the advantages of vehicle retarders:

....
....
....
....

Name three types of vehicle retarder:

....
....
....
....

Vehicle retarders usually operate on the transmission system to slow the vehicle and may be manual or automatic in operation; they can also provide varying degrees of retarding force.

Exhaust brake

The exhaust brake is a popular type of vehicle retarder; complete the drawing below to show the arrangement and describe briefly the principle of operation. Include on the drawing the driver's control and fuel cut-off.

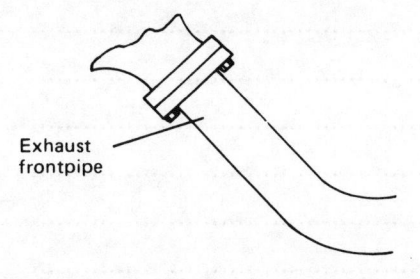

Exhaust
frontpipe

....
....
....
....
....
....
....
....
....

Name alternative types of shut-off valves:

....
....
....

ELECTRO-MAGNETIC RETARDER

The type of retarder shown is located in the transmission, usually directly behind the gearbox.

Name this type, label the drawing and describe briefly how it operates.

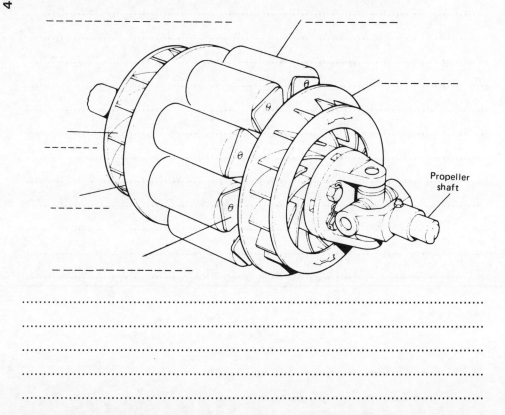

Propeller shaft

....

....

....

....

....

Make a simple line diagram to show how it is operated by the driver.

HYDRAULIC RETARDERS

The retarder shown below is usually situated in the driveline between the gearbox and final drive. In operation it is like a fluid coupling under 'stall' conditions, that is, the fluid is accelerated on to a stationary turbine (stator). The mechanical energy is therefore converted to heat energy. The degree of retardation is dependent on the amount of fluid in the retarder, this being controlled by a driver operated air-valve. On some vehicles the retarder is incorporated into the automatic gearbox.

Complete the drawing below to show the control system and heat exchanger.

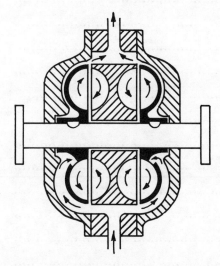

It is necessary to protect the braking system from hazards arising during repair or use.

List such hazards and protective measures necessary:

....

....

....

....

....

....

....

....

....

....

....

....

....

....

....

....

....

....

BRAKING SYSTEM MAINTENANCE

A preventive maintenance system will maximise braking system reliability, efficiency, service life and vehicle safety.

List preventive maintenance/MOT checks and tasks associated with the braking system.

Check:

....

....

....

....

....

....

....

....

....

....

....

....

....

....

....

....

Describe any special equipment used for bleeding a hydraulic braking system:

....
....
....
....
....
....

In relation to the braking system, for what purpose would the following equipment be used:

HYGROMETER ....

....

MULTIMETER ....

....

State the reason for and describe the process of skimming brake drums and discs:

....
....
....
....
....
....

List the general rules for efficiency and any special precautions to be observed when maintaining and repairing braking systems:

....
....
....
....
....
....
....
....
....
....
....
....
....
....
....
....
....
....
....

BRAKING SYSTEM FAULTS, SYMPTOMS, PROBABLE CAUSES AND CORRECTIVE ACTION

Complete the table in respect of the faults listed.

FAULTS	SYMPTOMS	PROBABLE CAUSES	CORRECTIVE ACTION
Worn friction material			
Seized wheel cylinder pistons			
Oil on brake lining			
Incorrect adjustment			
Air in system			
Fluid leakage			
Seized parking brake cable			
Faulty ABS sensor			
Worn brake drum or disc			

Chapter 10

Bodywork

29.1 VEHICLE BODY

Factors affecting body design

The intended use of a vehicle and cost constraints are two main factors affecting the design of a vehicle body.

Size and basic dimensions

Show and name the main dimensions on the body outlines below.

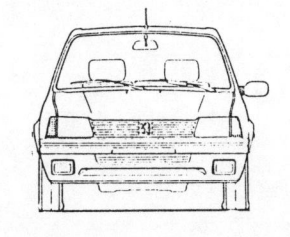

Volume, shape and location of space available for passengers and loads are vital factors.

Add the important passenger and load dimensions to the drawing below.

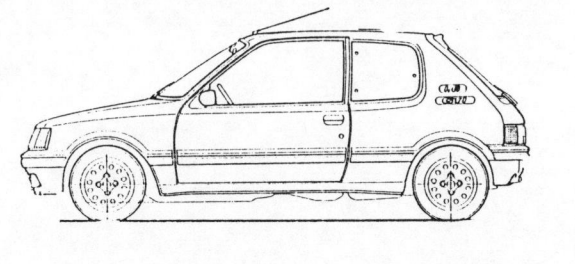

Explain the difference between GROSS VEHICLE WEIGHT (GVW), PAYLOAD and CAPACITY in relation to the vehicle and body.

GVW ....
....

PAYLOAD ....
....

CAPACITY ....
....

Strength

What forces is a body subjected to during normal operation?

....
....
....
....

What do the shaded and white areas represent on the vehicle shown below?

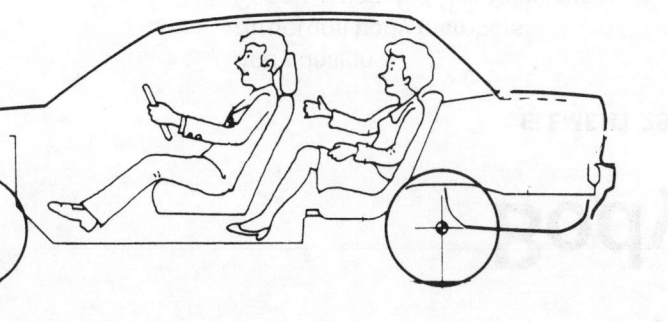

....
....

SHAPE OF BODY MEMBERS

Cross-sectional shape of key structural body members is an important factor with regard to body strength and rigidity.

Examine various body sections (or refer to a manual) and make sketches below to illustrate the cross-sectional shape of the body members listed.

INNER AND OUTER SILL

DOOR PILLAR

WINDOW PILLAR

CANTRAIL

CHASSIS MEMBER

FLOOR SECTION

State the differences between a STRESSED body panel and an UNSTRESSED body panel, and give two examples of each:

....

....

....

....

....

Give examples of where on a vehicle body the following methods of joining/attachment are used:

Welding ....

Adhesive joining ....

Mechanical fixing ....

Resilient (rubber) mountings ....

....

The CENTRE OF GRAVITY is the centre of the entire mass of the vehicle. What major factors determine the position of the C of G and how does its position affect vehicle operation?

....

....

....

....

....

....

See page 186 for more details of C of G.

CONSTRUCTION AND USE REGULATIONS

The purpose of the Construction and Use Regulations is to control the manner in which motor vehicles (and their equipment) are constructed, adapted and used. This is to ensure that legal standards regarding weight, construction and use, dimensions and design for vehicles and trailers are complied with. The Construction and Use Regulations do therefore contribute to the safety of vehicles on the road.

What is the purpose of the Type Approval Regulations?

....

....

....

....

....

....

....

....

List the areas of vehicle body construction and condition controlled by the Construction and Use Regulations:

....

....

....

....

....

....

Describe briefly how to remove and refit any TWO of the following and make simple sketches to illustrate the method of attachment, such as bolts, screws, clips etc.:

(a) subframe; (b) bonnet/boot lid/door/bumper/wing;
(b) trim/badges/rubbing strips/lights; (d) door seals; (e) seat/carpets;
(f) component access cover; (g) front/rear screen/quarter light.

VEHICLE MAKE/MODEL

....

....

....

....

....

....

....

....

....

....

....

....

....

....

....

....

....

....

29.3 Describe the procedure for installing a component on the body of a vehicle which involves drilling or cutting, for example, radio aerial, light etc. Sketch any special cutting tool used in the operation.

....

....

....

....

....

....

....

....

....

....

TOWBARS

A towbar is a bracket, usually 'T' or 'V' formation, which is attached to the rear underframe of a vehicle. The shape and attachment points of a towbar vary according to the vehicle to which it is being fitted. There is, however, a main design consideration for a towbar. What is this?

....

....

....

....

Make a sketch, top right, to show the installation of a towbar at the rear end of a vehicle. Include the body members to which the towbar is secured.

Towbar attachment

VEHICLE MAKE/MODEL ....

BODY MAINTENANCE

The benefits of routine maintenance and running adjustments on the vehicle body are:

Maintenance of appearance.

....

....

....

List routine maintenance tasks associated with the vehicle body:

....

....

....

....

....

....

....

....

Describe the procedure for:

1. Removing corrosive residues (salt/mud/tar etc.) from paint finishes and under bodies.

....

....

....

....

2. Touching up minor paintwork damage.

....

....

....

....

....

Methods used to protect the body/chassis against corrosion and damage arising from use or repair include the use of such as:

(a) galvanised panels (doors, bonnet, boot lid etc.)

(b) cathodic phosphate electrocoating

(c) multi-layer painting (primer, synthetic enamels etc.)

(d) wax injection

(e) stone chip protection

(f) flange sealers

(g) anti-scuff strips

(h) plastic wing liners

(i) transit protective coating

Indicate on the drawing, top right, where the protective measures listed would be likely to be applied.

How is the protection maintained during vehicle use and repair?

1. ....

....

....

....

....

....

2. ....

....

....

....

....

Chapter 11

Electrical and Electronic Systems

ELEMENT 28   **UNITS 43/44**

Note: In accordance with City and Guilds requirements for the 383 course, for additional work in this area also see Chapter 13 of Engines and Related Systems.

BASIC ELECTRICAL CIRCUITS

The electrical/electronic system in a motor vehicle is very complex, therefore to allow an understanding of how the system works, it must be broken down into individual circuits. The individual electrical circuits fitted to light, heavy and public service vehicles may use, in some form, the items listed in the table below. For each of the items listed state its functional requirement and basic working principle.

ITEM	FUNCTIONAL REQUIREMENT	WORKING PRINCIPLE TO ACHIEVE FUNCTIONAL REQUIREMENT
Generation and storage		
Motor and drive assemblies		
Lights		
Heating elements		
Switches and relays		
Conductors and terminations		
Circuit protection devices		
Driver information circuits		

28.5 SYMBOLS

Sketch the appropriate symbol which is normally used to represent each of the components listed.

BATTERY		DIODE		GAUGE	
Switch		Zenor diode		Horn	
Relay		Thermistor		Alternator	
Fuse		Thyristor		Starter motor	
Bulb		Plug and socket connector		Voltmeter	
Motor		Radio		Ammeter	
Resistance		Speaker		Ohmeter	
Capacitor		Transistor		Variable resistor	

28.5 LEAD – ACID BATTERY

The lead – acid battery is used on most automobiles in either 12 or 24-V form.

A 12-V battery container consists of six separate compartments. Each compartment contains a set of positive and negative plates; each set is fixed to a bar which rises to form the positive or negative terminal. The plates have a lattice-type framework into which is pressed the chemically-active material. Between each plate is an insulating separator.

Name the parts on the sketch below.

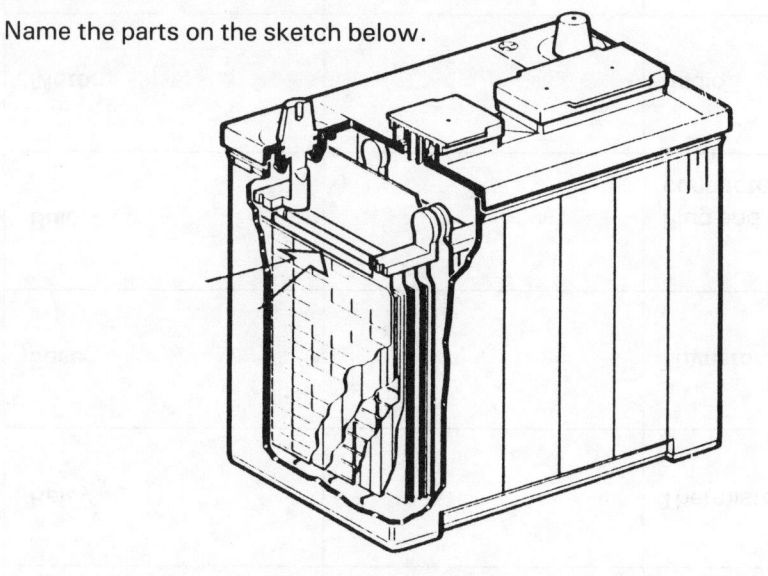

Since each cell has a nominal electrical pressure of 2 V, to produce a 12-V battery six cells must be joined together in series.

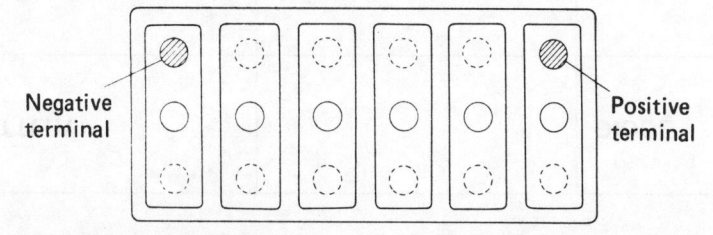

Negative terminal

Positive terminal

The cells are connected by buss bars (or links), not normally seen on modern batteries. Show where these would be connected in the diagram above and indicate the polarity of each cell by symbols.

Modern batteries may be classed as low-maintenance or maintenance-free batteries.

What feature allows no maintenance except for keeping the battery case and top clean and dry?

....

....

....

....

BATTERY CLASSIFICATION AND RATING

The basic size of a battery gives some indication as to its potential output. Below are shown typical specifications for two 12-V low maintenance batteries. Explain what the stated terms mean.

	SPECIFICATIONS	CAR	hgv
A	Cold start performance (amps)	225	550
B	Reserve capacity at 25 amps (mins)	85	220
C	Bench charge rate (amps)	5.5	12.5
D	Capacity (20 hour rate) amp/hours	55	125

A ....

....

....

B ....

....

C ....

....

....

D ....

....

....

BATTERY CONDITION

The two principal features indicating the condition of a battery are:

1. State of charge

2. Ability to supply a heavy current for a short period of time.

What can be tested using a hydrometer?

....

What does the test indicate?

....

The hydrometer readings of a battery electrolyte will vary between about

.... and ....

Complete the table below to show a specific gravity range for the conditions listed.

State of charge	Specific gravity
Fully charged	
Half charged	
Discharged	

Name the main parts of the hydrometer shown:

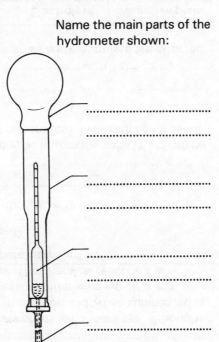

....

....

....

....

....

....

Complete the drawings below to show the expected float level position:

1.140 1.200 1.230 1.260 1.290

Discharged | One-quarter charged | Half charged | Three-quarters charged | Charged

What is meant by relative density (or specific gravity)? ....

....

....

Why do the relative density values indicate the state of charge of a battery?

....

....

....

....

....

Check the state of charge of a battery:

Take hydrometer and voltmeter readings of the battery.

Cell number	1	2	3	4	5	6
Relative density readings						
Voltage readings						

General condition of battery ....

....

What can be checked by a high-rate discharge tester?

....

State briefly how it should be used:

....

....

....

221

28.5 BATTERY CHARGING

The battery is a series of chemical cells. Each cell of a lead – acid battery is capable of producing 2 V. The size and number of plates in the cell determine its capacity or output. It is of a secondary cell type.

What does 'secondary cell' mean?

....

What is a primary cell? Give an example:

....

....

The positive and negative plate active materials of the secondary cell, although of a similar base, have a different chemical composition and when submerged in a suitable electrolyte produce an electrical pressure difference which when connected to a circuit allows a current to flow. The chemical reaction then discharges the cell until the plates become chemically similar.

Why is the cell then capable of being recharged?

....

....

....

When the battery requires charging, the supply, be it from the vehicle's charging system or the mains supply, must be a direct current (dc). Why must only direct current be used?

....

....

Complete the table to show how the plate materials and electrolyte are affected during the charge and discharge cycle.

Process	Positive plate	Electrolyte	Negative plate
Fully charged			
Discharged			

Name the two gases given off from the cells when the battery is being recharged:

1. .... 2. ....

When batteries are charged on a vehicle by the alternator, they are initially (after engine starting) subjected to a high rate of charge and then controlled to a much lower charge rate. When removed from the vehicle and externally charged, a fast or slow charger may be used; it is preferable to .... charge if time allows.

What major precautions should be taken if an external charger is used to charge a battery still connected to the vehicle?

(1) Connect a battery to an external charging system.
    What way (electrically) are the charger-to-battery connections made?

....

(2) Connect three 12-V batteries to a constant current charger. Show on the sketch how the batteries are connected to the charger.

The cells are connected in.............

Constant current battery charger

List the precautions required when charging batteries.

....

....

....

28.5 THE ALTERNATOR

In order to produce an electric current by magnetic induction, three basic requirements must be fulfilled.

In an alternator these are:

....

....

....

....

....

....

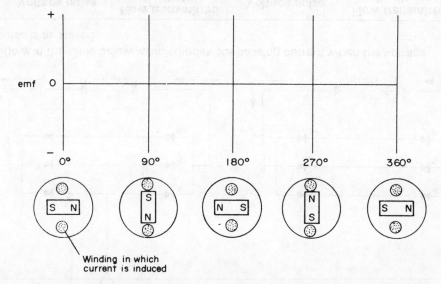

Label the basic requirements on the simple sketch above.

Describe the two main parts of the alternator shown below.

....

....

....

....

Action of the alternator

As the rotor turns inside the stator, the magnetism creates a pulse of current in the stator winding. Show the emf wave form produced by one complete turn of the rotor (magnet) and indicate on the drawings below the direction of flow of the magnetic field for each rotor position.

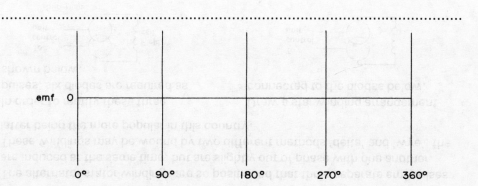

Winding in which current is induced

The windings in an alternator stator frame consist of three sets of wires wound across each other (shown opposite). State the reason for this and show below the wave form produced.

....

....

223

28.5 ALTERNATOR — PRINCIPLE OF FULL-WAVE STATIC RECTIFICATION

When the magnet in a simple alternator is revolved, one complete turn, an emf is induced in the circuit, first in one direction and then in the reverse direction.

On the diagrams below show an alternating pulse and a rectified pulse.

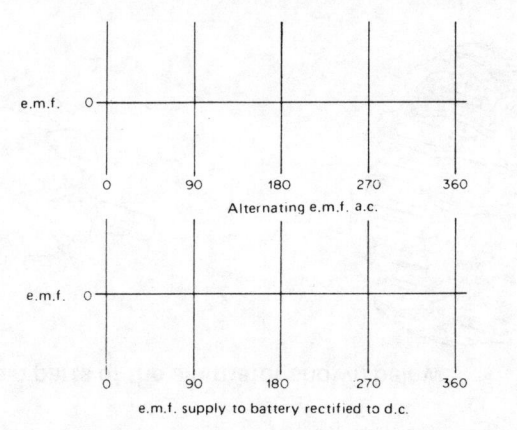

Alternating e.m.f. a.c.

e.m.f. supply to battery rectified to d.c.

The components that allow this rectification to occur are diodes (see page 226).

Show, using arrows, how diodes rectify the supply of current induced in a single coil of wire.

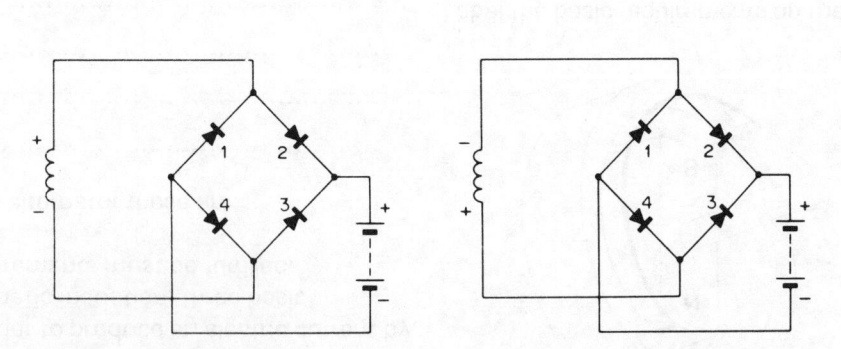

Diodes passing current nos ....   Diodes passing current nos ....

The alternator stator windings are so positioned that three separate emf pulses are induced at the same time, but are slightly out of phase with one another. These windings may be wound by two different methods 'delta' and 'wye', the latter being the more popular in this country.

In order to rectify these three pulses, six diodes are required as shown below.

Draw a star winding arrangement connected to the diodes below.

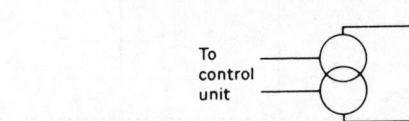

Delta winding

Star or Wye winding

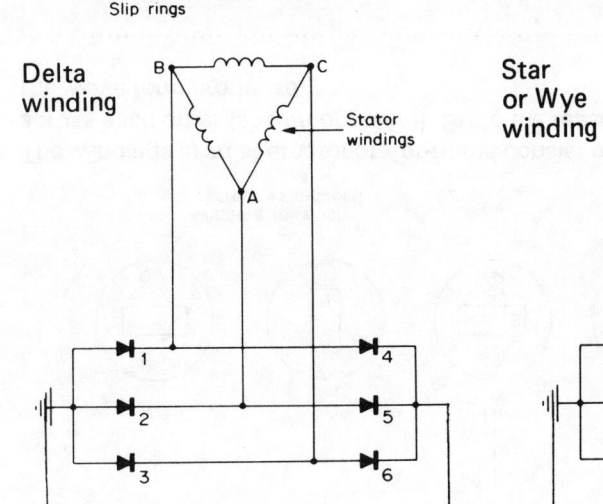

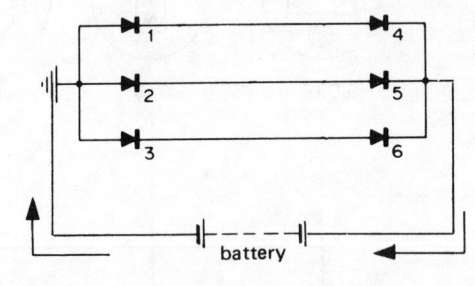

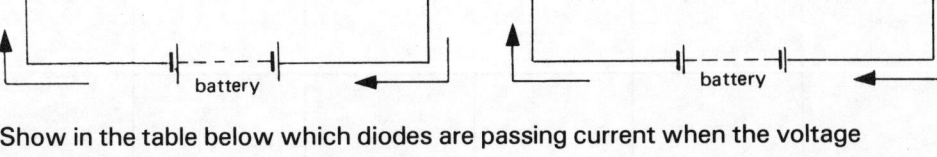

Show in the table below which diodes are passing current when the voltage pulse is as stated.

Voltage pulse	Flow transmitted by diodes	Voltage pulse	Flow transmitted by diodes
A to B	4 and 2	C to A	
B to A		B to C	
A to C		C to B	

28.5 THE ALTERNATOR

The alternator shown opposite consists of a series of magnets which are rotated in the centre of three sets of inter-wound coils.

Lucas
AC type

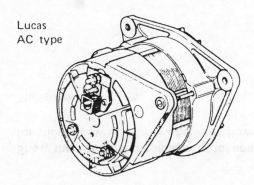

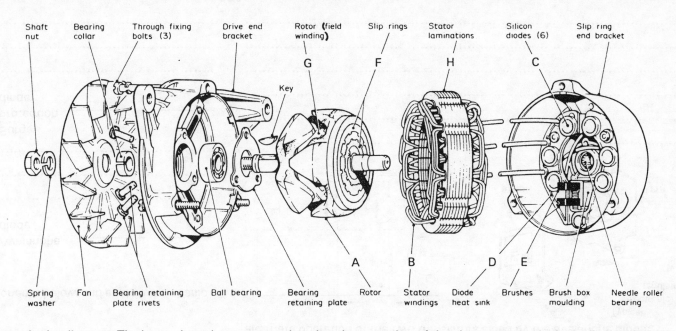

Shaft nut • Bearing collar • Through fixing bolts (3) • Drive end bracket • Rotor (field winding) • Slip rings • Stator laminations • Silicon diodes (6) • Slip ring end bracket

Key

G   F   H   C

A   B   D   E

Spring washer • Fan • Bearing retaining plate rivets • Ball bearing • Bearing retaining plate • Rotor • Stator windings • Diode heat sink • Brushes • Brush box moulding • Needle roller bearing

Explain the function of the lettered items shown in the diagram. The letters have been arranged so that the operation of the alternator may be logically developed.

A ....

....

....

B ....

....

....

C ....

....

....

D ....

....

E ....

....

....

F ....

....

....

G ....

....

....

H ....

....

Note: For other types of alternator, see Chapter 15 in _Engines and Related Systems_.

FUNCTIONS OF SEMI-CONDUCTOR COMPONENTS

The electronic components listed below are all used in the alternator's electrical circuit and may be used in any other circuit that adopts electronic control.

The alternating current flow must be rectified to direct current. This is achieved by using a number of static rectifiers called diodes.

The function of a diode is to:

....

....

Show a diode's electrical symbol and indicate the direction of current flow.

What material is used in order to allow this to occur? ....

....

....

Show the electrical circuit symbol for components below and explain their basic function. Show directions of current flow.

Transistor

Avalanche diode   _____ _ _____

....

....

....

....

Surge protection diode   _____ _ _____

....

....

....

....

ELECTRONIC REGULATOR

The charging rate of alternators may be controlled by either a mechanical vibrating point voltage regulator, or by an electronic regulator.

Why does the charge output of the alternator require control?

....

....

....

How is the alternator's output basically controlled?

....

....

....

Below is shown a simplified circuit diagram of an electronic regulator. With the aid of the diagram, explain how the emf voltage build-up is controlled by the alternate operation of the two transistors aided by the avalanche diode.

Alternator

Typical electronic regulators

LUCAS 12V

8TRD

11TR

....

....

....

....

STARTER MOTOR

Starter motors fitted to cars are usually one of the two types shown. What is the purpose of the starter motor?

....

....

Examine starter motors of the types shown and identify the arrowed parts.

Type (I)

....

....

Type (II)

....

What is the function of the solenoid switch?

....

....

....

....

The diagram opposite shows the basic components in a simple starting system.

Sketch in the appropriate wiring, indicate which components need to be earthed and distinguish between the different types of cables.

Explain the mechanical operation of:

Type (I) ....

....

....

....

....

....

....

....

....

Motor armature

Type (ii) ....

....

....

....

....

....

....

....

....

....

Name the parts indicated.

Battery

Solenoid switch

M

Ignition switch

Solenoid connection

227

MOTOR OPERATION

Describe, with the aid of the simple diagrams below, the basic operation of an electric motor (such as a starter motor).

Armature winding

....
....
....
....
....
....
....
....

Windings increased

....
....
....
....
....
....
....
....

Field windings connected in parallel

....
....
....
....

Examine a starter armature and count the number of segments on an actual commutator.

No. of SEGMENTS ....

No. of LOOPS ....

STARTER MOTOR WIRING

Field windings can be wired in a number of ways. Complete the diagrams to show the basic theoretical methods listed below.

Two types of field coil
may be used:

Series ....

....

....

Parallel ....

....

....

SERIES MOTOR

What factors determine the
type of design used?

....

....

....

....

SHUNT MOTOR

State the characteristics possessed by a:
Series wound starter motor

....

....

....

....

COMPOUND MOTOR

Parallel (shunt) wound starter motor

....

....

....

....

Where else on a vehicle might similar type motors be employed?

....

....

ROUTINE MAINTENANCE

State FOUR main reasons for carrying out routine preventive maintenance on the vehicle electrical system:

1. ....

2. ....

3. ....

4. ....

What is the first check that should be carried out on the electrical system?

....

....

List FOUR basic checks that should be carried out during a normal service.

1. ....

2. ....

3. ....

4. ....

Describe how to tension a drive belt:

....

....

....

....

On non-maintenance-free batteries, what liquid should be used to top-up the electrolyte?

....

When should topping-up be required and what quantity should be used?

....

....

What is the maintenance requirement for battery terminals?

....

....

....

Describe how the electrical systems components can be protected during use or repair from hazards such as:

Incorrect polarity

....

....

Electrical overload

....

....

Contamination by dirt and moisture

....

....

Excessive heat

....

....

Precautions

List precautions to be observed when carrying out routine electrical maintenance and repair in terms of:

Checks and tests

....

....

....

....

....

....

....

Batteries

....

....

....

....

CABLES

Components within an electrical circuit are interconnected by flexible copper cables which provide a path for the electrical current. When selecting a cable for a particular application, a number of factors must be considered, such as current to be carried, length of cable required, routing protection/insulation, identification etc.

Cable size and current rating

The size of a cable is designated by the number of strands of wire and the thickness of each strand.

A typical cable size used on vehicles is ....

The first number represents ....
and the second number is ....

The actual 'current rating' for such a cable is ....
....

The size of cable is increased as the current carrying requirement increases. Why is this necessary?

....

....

....

What influence does the length of cable have on its size? ....
....

Complete the table below to give the cable sizes and ratings for the applications listed.

APPLICATION	RATING (AMPS)	SIZE
Side and tail lamps		
Head lamps		
Alternator		

What materials are used for cable insulation?

....

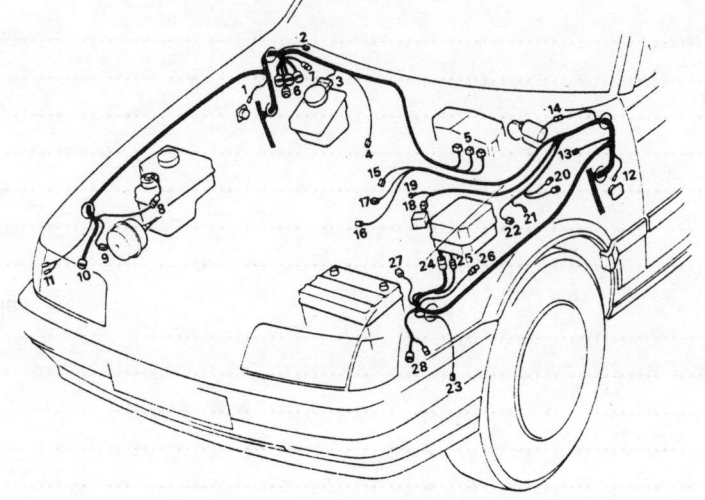

The illustration above shows a main engine compartment harness or loom. A number of wiring harnesses are used throughout a vehicle to bind together wires servicing a common area. Why is this system used?

....

....

Colour coding

Complete the table.

CODE	COLOUR	FUNCTION
B		
G		
N		
R		
U		
W		
P		

Why are the code letters used?

....

TERMINALS AND CONNECTORS

Terminals and connectors are used to attach cables to components, switches etc, and to interconnect different parts of circuits.

Identify the following types of connectors:

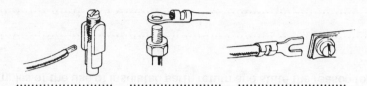

....

Sketch the male half of the snap connectors 1 and 2:

....

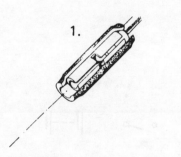

1.

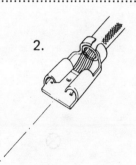

2.

Examine main harness connectors such as the type shown below and made simple sketches on the right to illustrate (1) the connector locking device, (2) the method of securing cables in the connector.

CIRCUIT PROTECTION

State the function of a fuse or circuit breaker:

....

....

The glass type fuse shown at (a) contains a paper indicating the amperage rating. The more recent type (b) is colour coded for rating purposes. Name the fuse types at (c) and (d).

(a)   (b)   (c)   (d)

....

Explain the difference between PEAK CURRENT rating and CONTINUOUS RATING of fuses:

....

....

How does a FUSIBLE LINK compare with the fuses shown and where would it be used?

....

Bi-metal circuit breakers, in which the heating effect of excessive current would bend the bi-metal strip and cause contacts to break, is another form of circuit breaker. These are very often the 'manual reset' type. Describe this method of circuit breaking and give an example of its use.

....

....

....

....

28.5 CIRCUITS

The electrical circuits in this book are generally, as in most vehicles, EARTH
RETURN in which the earth path for the circuit is provided by the body/chassis.
This means that the current passes from the battery through switchgear to a
component, such as a light bulb, and then through the chassis frame back to the
battery to complete the circuit.

Complete the drawings below to show (a) an EARTH RETURN circuit and (b) an
INSULATED EARTH RETURN circuit.

(a)

$$I|---I| \qquad \circ\!\!\!\!\!-\!\!\!\!\circ \qquad \otimes$$

(b)

$$I|---I| \qquad \circ\!\!\!\!\!-\!\!\!\!\circ \qquad \otimes$$

Give examples of the use of insulated earth return and state the reason for its
use:

....

....

....

....

Complete the electrical circuits opposite by adding the wiring layout and
components such as: switches, relays, circuit breakers, resistors etc.

Include on the drawings the colour coding for the circuits.

Charging circuit

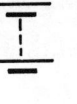

Heated rear-window circuit

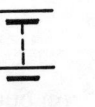

BULBS AND LAMPS

The drawings opposite illustrate typical front and rear lamp assemblies.
Complete the labelling on the drawings and the table below.

BULB	TYPE/RATING DESCRIPTION	FUNCTION
A		
B		
C		
D		
E		

How does the SEALED BEAM headlamp unit differ from the type shown opposite?

....

....

....

Lenses

The lense in for example a headlamp distributes the light rays to provide correct illumination of the road ahead during main beam and dip operation.

State the purpose of the area (x) on the lens shown below.

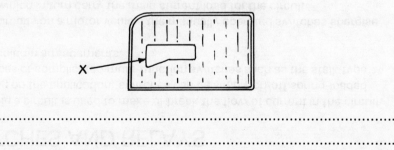

....

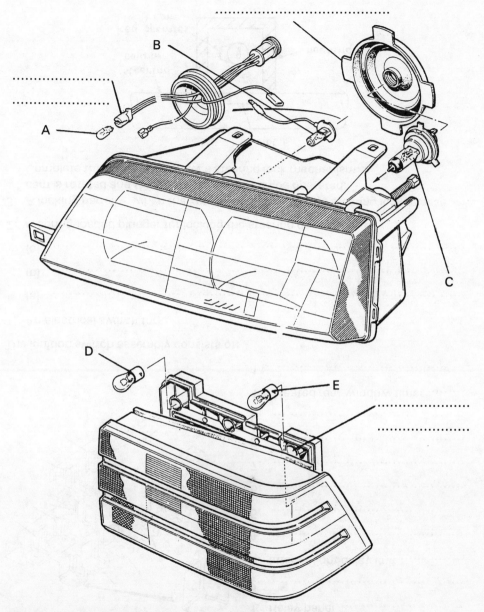

Show how the lens opposite can be modified for continental driving.

233

SWITCHES AND RELAYS

A switch in a circuit is used to make or break the flow of current in the circuit. Depending on the application, switches can be simple on/off spring-loaded toggle types or complicated multi-function switches such as the stalk type steering column arrangements.

In many circuits on a motor vehicle the manually operated switches energise RELAYS which in turn carry the main current load for the circuit.

Give two reasons for the use of relays:

....

....

Relay operation (Note: Also see horn section)

A typical relay panel is shown opposite.

Complete the list opposite with names of typical relays.

The interior light circuit employs a DELAY relay and the heated rear window employs a TIMER relay. Describe how these relays operate within their circuits:

....

....

....

....

....

....

....

....

....

Give an example of an INTERMITTENT RELAY:

....

....

1. Relay panel ....

2. ....

3. Interior light delay unit ....

4. ....

5. ....

6. ....

7. ....

8. ....

9. ....

10. Heated rear-window timer ....

11. ....

The ignition switch assembly consists of:

1. An electrical switch for:

(a) ....

(b) ....

(c) ....

2. A spring-loaded plunger for locking the steering column.

3. A locking barrel — when the key aligns the wards the steering lock release cam is rotated and the electrical switch can be operated.
   Complete the drawing to show a steering lock mechanism.

Chapter 12

Lighting Systems

LIGHTING

The purpose of the lighting system on a vehicle is to:

(a) Identify the vehicle and indicate its position.

(b) ....

(c) ....

(d) ....

(e) ....

State the main statutory regulations relating to:

(a) Bulb rating

....

....

(b) Colour of light emission

....

....

....

(c) Lens and reflector condition

....

....

(d) Fog and driving lamps

....

....

(e) Reversing lamps

....

....

(f) Use of lamps

....

....

Lamps and reflectors fitted to road vehicle lighting systems include:

1. .... 7. ....

2. .... 8. ....

3. .... 9. ....

4. .... 10. ....

5. .... 11. ....

6. .... 12. ....

Complete the wiring diagrams for the circuits 1 to 5; include any fuses, relays etc. in the circuits.

1. Side and tail lamp circuit

2. Headlamp circuit

3. Fog and driving lamps

The fog lamp gives a low flat beam to pick out the kerb, whereas a driving lamp provides a long penetrating beam useful for fast driving.

4. Reversing lamps

5. Interior lights

PSV INTERIOR LIGHTING

The interior lighting load for a PSV is obviously considerable. It is therefore advantageous to employ fluorescent lighting for this purpose. State two advantages of fluorescent lighting:

1. ....

2. ....

It is common practice to use three-foot fluorescent tubes operating in pairs in conjunction with an inverter and transformer. What is the purpose of these two components?

....

....

....

....

....

HEADLAMP ALIGNMENT

All vehicle headlamps in the UK must comply with the Department of Transport Road Vehicle's Headlamp Regulations, which state the position, or angle, of the dipped beams.

The specialist equipment used to align headlamps measures the angle of dipped beam and the beams' positions relative to one another when both on main and dipped beam.

Show a sketch of such equipment in its testing position.

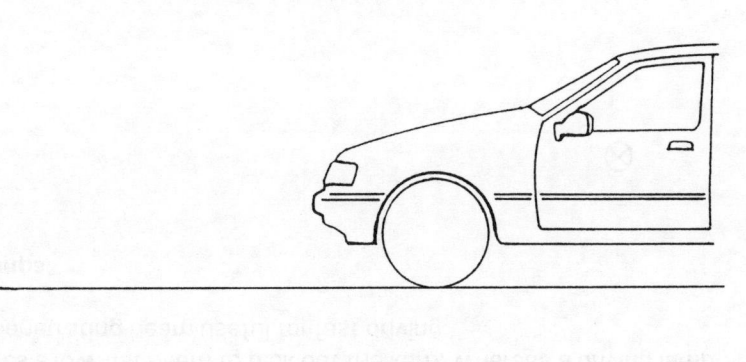

Type of gauge:

What pre-checks are necessary to ensure that headlamp alignment is accurately carried out?

INVESTIGATION

Check the headlamp alignment of a vehicle, using available equipment, and describe the alignment procedure.

Vehicle Make .... Model ....

....

....

....

....

On the aiming screens shown, indicate correct main and dipped beam positions.

1. Typical main driving beam position all types

....

....

....

....

....

....

....

Vertical screen | Aiming line

Horizontal screen

Aiming line

2. Symmetric dipped (passing)beam

....

....

....

....

....

....

....

3. Asymmetric dipped (passing) beam

....

....

....

....

....

....

DIRECTION INDICATORS/HAZARD WARNING

Amber coloured flashing lights are used to signal a hazard warning, or intent to change the direction of a vehicle, to other road users. In the indicator circuit a FLASHER UNIT is used to make and break the electrical supply to the lamps causing them to flash.

Complete the wiring diagram for the circuit shown and describe the operation of the system during normal turn indicating and hazard warning.

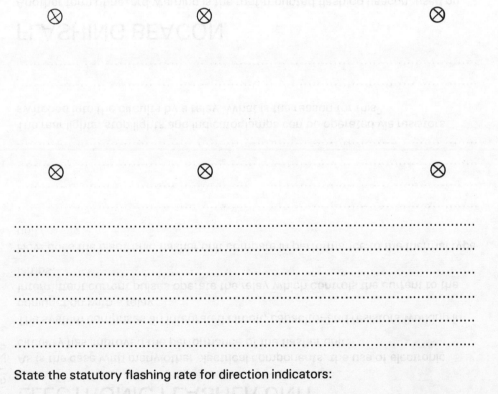

....

....

....

....

....

....

....

State the statutory flashing rate for direction indicators:

....

FLASHER UNIT (THERMAL TYPE)

The VANE type flasher unit shown below relies on the heating effect of the electric current on a thin metal RIBBON to provide intermittent switching of the flasher lamp circuit.

The base supports a snap-action metal vane held in tension by a metal ribbon.

When not in operation, the contacts are closed.

Name the main parts, and indicate current flow.

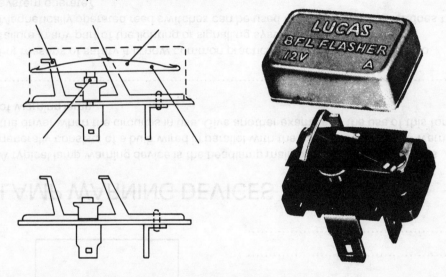

Show position of ribbon and vane when contacts are open.

Describe its operation:

....

....

....

....

ELECTRONIC FLASHER UNIT

As is the case with many other electrical components, the use of electronic circuitry has improved the performance of the flasher unit.

The system comprises an integrated circuit, capacitor and resistors working in conjunction with a relay.
Intermittent current pulses operate the relay which controls the current to the lamps.

How does the electronic flasher unit compare in performance to the thermal type unit?

....
....
....
....

The rear lights, stop lights and indicator lamps can be operated via resistors switched into the circuits by a relay. What is the reason for this?

....
....

FLASHING BEACON

Another form of hazard warning is the roof-mounted flashing beacon, used on breakdown vehicles etc. This type of lamp can be operated in a similar way to the direction indicators or the flashing can be achieved by a rotating reflector. Examine a flashing rotating beacon, complete the drawing opposite and describe briefly how it operates.

State the statutory regulations relating to flashing beacons:

....
....
....
....
....

....
....
....
....
....
....
....

LAMP WARNING DEVICES

A typical lamp warning device is the headlamp main beam warning light. This generally consists of a bulb wired in parallel with the main beam lamps to warn the driver when the circuit is in use. Give another example of the use of this form of warning lamp.

....

For reasons of safety it is now common practice to warn the driver of a bulb failure in any part of the lighting or signalling system.
Magnetically operated reed switches can be used for this purpose. How does the system operate?

....
....
....
....

Give an example of the use of a 'switch off reminder':

....

TRAILER/CARAVAN ELECTRICS

All types of trailer units (including caravans) must show the obligatory rear-facing lamps. These must be operated from the driving vehicle's battery and not a separate battery (that is, that could be fitted to the trailer to provide interior lighting).

In order to provide this supply safely, and allow it to be easily disconnected from the vehicle, 7-pin connectors are used. An example is shown below.

Identify each part.

 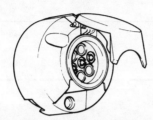

On older vehicles and caravans a single 7-pin connector is able to supply all the electrical needs. On modern caravans it is obligatory to fit rear fog lights. This extra item, together with the possible fitting of reversing lights and internal caravan electrical equipment, has made the fitting of two 7-pin connectors essential.

The first (or existing) connector is known as a ....

The second connector is known as a ....

Number each connection on the drawings and indicate how the sockets are arranged to prevent interchangeability.

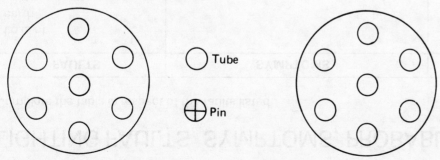

The vehicle's rear cable harness arrangement shows the items that should be connected to the 12 N socket. With the aid of the table below, complete the diagram to show the socket correctly wired.

Number the socket connections and indicate the cable colours.

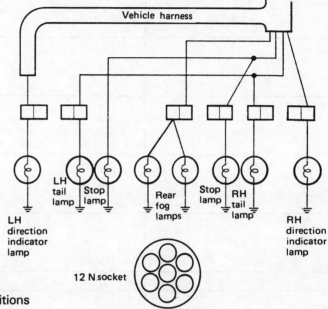

Pin dispositions

Pin. no.	7-core cable colour	Circuit allocation	
		12 N connector (ISO 1724)	12 S connector (ISO 3732)
1	Yellow	LH flashers	Reverse and/or mechanisms
2	Blue	Rear fog (auxiliary, older vehicles)	No allocation (additional power)
3	White	Common return (earth)	Common return (earth)
4	Green	RH flashers	Power supply (caravan interior)
5	Brown	RH side/tail/no. plate	Sensing device (warning light)
6	Red	Stop	Power supply (refrigerator)
7	Black	LH side/tail/no. plate	No allocation

LIGHTING FAULTS, SYMPTOMS, PROBABLE CAUSES AND CORRECTIVE ACTION

Complete the table in respect of the faults listed.

FAULTS	SYMPTOMS	PROBABLE CAUSES	CORRECTIVE ACTION
Loss of earth			
Corroded bulb holder			
Beam misalignment			
Physical damage			
Faulty flasher unit			
Corroded or discoloured lamp reflector			
Short circuit			
Faulty switch			
Open circuit			

VOLT DROP CHECK

When checking for high-resistance (bad connections), a volt-drop check should be made.

State what is being checked on the diagrams and carry out a similar check.

Give expected and actual values.

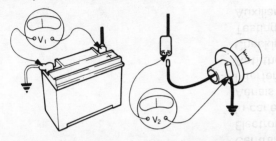

Readings	Voltage Expected	Actual
V_1		
V_2		
V_3		
V_4		
V_5		

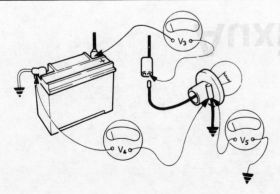

CURRENT CONSUMPTION

Describe how the wiper motor circuit shown would be checked for current consumption. Show on the circuit where the instrument would be connected.

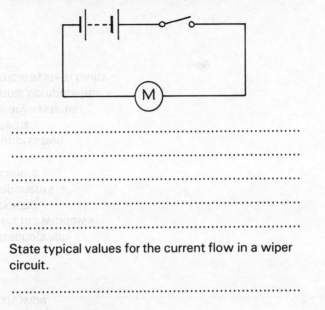

....

....

....

....

....

State typical values for the current flow in a wiper circuit.

....

OHMETER

State one example of the use of an ohmeter and show by sketching how the meter could be connected to check circuitry or components:

....

....

What are the advantages of a MULTIMETER as opposed to the use of individual gauges?

....

....

....

....

....

Calculate the current consumed by a 12 V 21 W lamp bulb.

....

....

....

....

....

....

The current supplied to a hgv wiper motor during operation is 25 A and the voltage at the motor terminals is 12.5 V. Calculate the power consumed by the wiper.

....

....

....

....

....

Chapter 13

Auxiliary Electrical Systems

ELEMENT 53   **UNITS 43/44**

AUXILIARY ELECTRICAL SYSTEMS

Auxiliary electrical systems fitted to road vehicles include the following:

1. Windscreen and headlamp wipers and washers.

2. ....

3. ....

4. ....

5. ....

6. ....

7. ....

WINDSCREEN WIPER SYSTEM

Statutory regulations require that a road vehicle should be equipped with one or more windscreen wipers and washer to give the driver a good view of the road ahead in all weather and driving conditions. Most windscreen wiper mechanisms are electrically operated. A relatively powerful motor is required to drive the mechanism and it is desirable that the motor and mechanism are quiet in operation.

Describe the operational features of a windscreen wiper/wash system:

....

....

....

....

....

....

....

WINDSCREEN WIPERS

The wipers are operated by a mechanism which is given a to-and-fro action by a suitably designed linkage which is connected to the armature shaft of a small electric motor.

Identify the types of drive assembly.

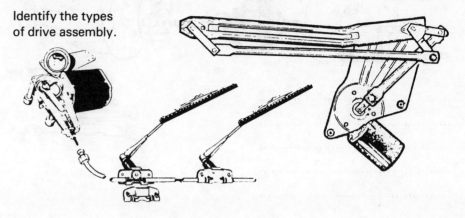

....

The electric motor may be of a single-twin or variable-speed type. Almost all types have a limit-switch device incorporated into the drive assembly.

Below is shown the wiring diagram of a single-speed windscreen wiper motor. Explain the operation of the limit-switch when the main control switch is opened.

....

....

....

....

....

....

....

SINGLE- AND TWO-SPEED WIPER MOTORS

The electric motors used for wiper operation are normally two-speed PERMANENT MAGNET type. How do permanent magnet motors differ from the WOUND FIELD type?

....

....

....

....

....

Describe how two speed motor operation is achieved. Complete the sketch below to show the brush layout and circuit for two-speed operation.

....

....

....

....

....

Describe the drive arrangement on both motors and state if single or twin speed, giving reason for choice.

....

....

....

....

....

....

....

Wiper is .... speed; there are .... brushes.

....

....

....

....

....

Wiper is .... speed; there are .... brushes.

INTERMITTENT WIPE SYSTEM AND SCREEN WASH

The key components in the wiper electrical circuit controlling the intermittent wipe operation are:

1. Wiper Relay   2. Park Switch   3. ECU.

The ECU provides a current pulse, for example 0.65 seconds every 6 seconds to energise the relay when intermittent wipe is selected.

Study the wiper circuit shown on the next page and describe the operation during intermittent wipe.

Describe also the operation during screen wash.

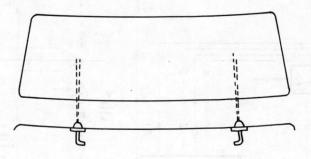

....

....

....

....

....

....

....

....

....

....

....

....

....

A THERMAL PROTECTION device can be fitted in series with the electrical supply to the wiper motor. State the function of such a device and briefly describe one type:

....

....

....

....

....

Examine a vehicle and complete the drawing top right to show the screen wash system layout.

....

HEADLAMP WASH/WIPE

On the more expensive type of vehicles, headlamps are often fitted with a
POWER WASH SYSTEM or a WASH/WIPE SYSTEM. The wash pump for the
headlamps is normally a higher powered pump than the screen wash pump and
is actuated by a timer.

....

....

....

....

....

....

....

Examine a vehicle and complete the drawing below to show the headlamp
wash/wipe arrangement.

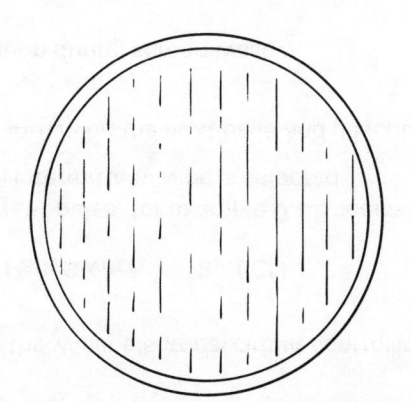

Circuit Diagram - Front wipers

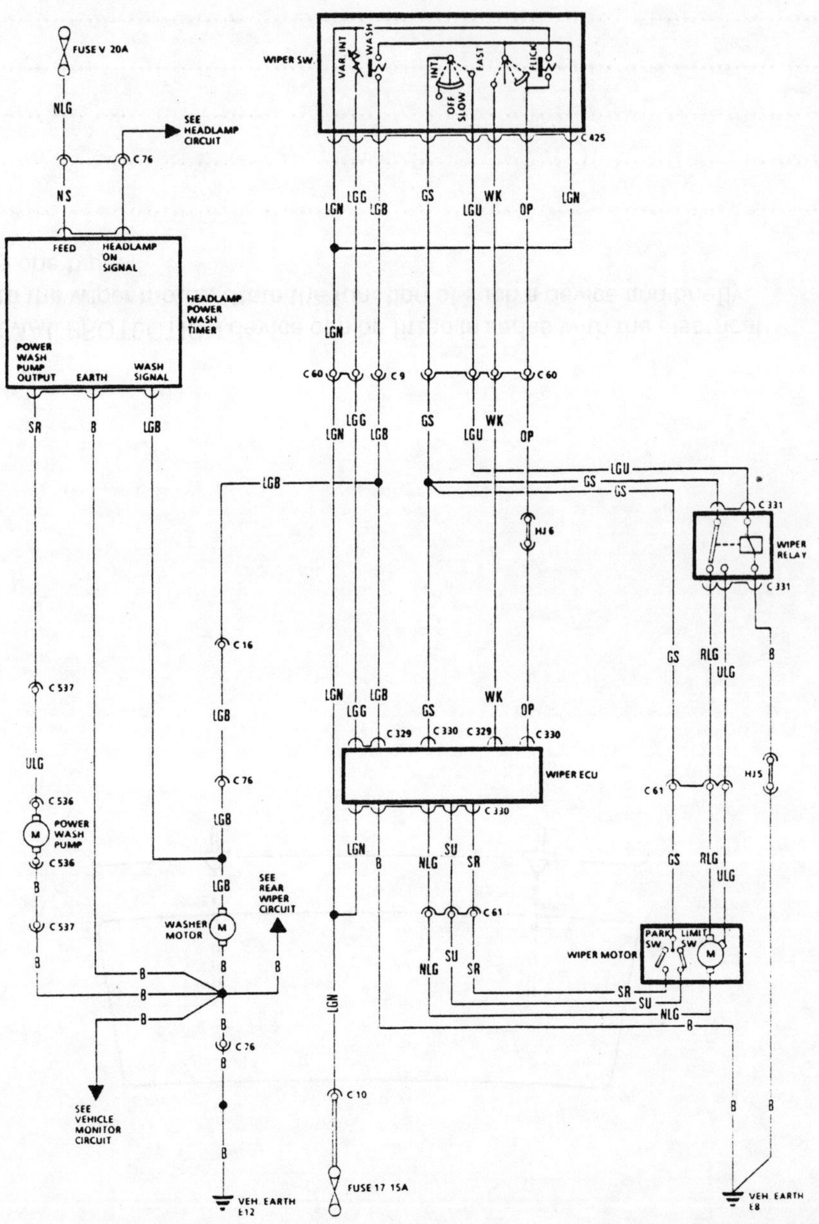

HORNS

A requirement of the Construction and Use Regulations is that all vehicles must be equipped with an audible warning device.

Three types of horn used for this purpose are:

1. *High frequency horn*

   ....

2. ....

3. ....

Complete the drawing below to show a 'high frequency' horn and explain how it operates.

Name the various parts

....

....

....

....

....

....

....

....

....

....

....

Terminals

Windtone horns shown opposite are often used in pairs on a vehicle. Why is this?

....

....

....

Complete the drawing below to show a 'windtone' horn and describe its operation.

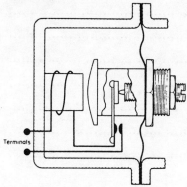

Terminals

....

....

....

....

....

Complete the wiring diagram of the simple twin-horn circuit shown below.

from battery   horn push

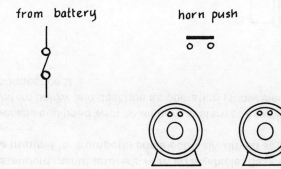

twin horns

RELAY SWITCHES

As in many electrical circuits on a vehicle, a relay is used in the horn circuit. Why is this?

....

....

....

Complete the wiring of the three-terminal relay shown and explain its principle of operation.

Indicate terminal connections.

....
....
....
....
....
....
....
....

Sketch a wiring diagram of a twin-horn circuit to include a relay switch.

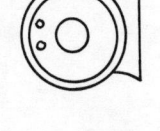

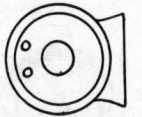

AIR HORNS

Air horns are usually purchased from a motor accessory shop and fitted to replace the standard manufacturer's horn on a vehicle. Basically the system consists of a trumpet (or trumpets) and electrically driven air pump.

Examine a vehicle equipped with an air horn system or examine a kit. Sketch the complete system below and describe its operation (show electrical and pneumatic connections).

....
....
....
....
....
....
....

BI-METAL FUEL AND TEMPERATURE GAUGES

These gauges are fitted on most modern vehicles. The current supply to both fuel and temperature gauge is controlled by a voltage stabiliser. The sender units in both fuel tank and cooling system have resistors that vary depending on either the amount of fuel in the tank or the engine temperature.

Explain the principle of operation of the:

Voltage stabiliser

....
....
....
....

Bi-metal gauge

....
....
....
....

Engine cooling system temperature transmitter

....
....
....
....

Complete this wiring diagram to show the internal wiring symbols of the voltage stabiliser, gauge units and temperature and fuel gauge transmitter units.

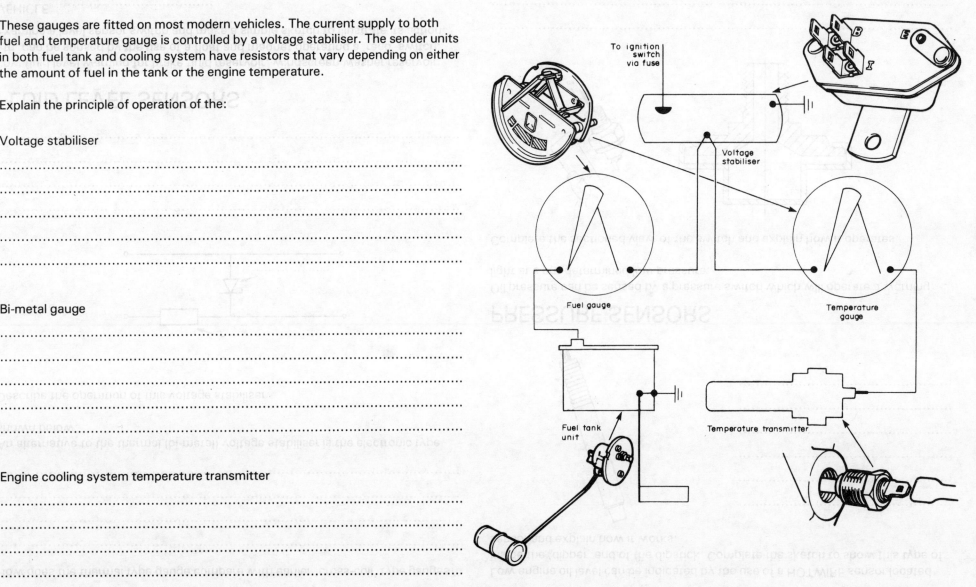

To ignition switch via fuse

Voltage stabiliser

Fuel gauge

Temperature gauge

Fuel tank unit

Temperature transmitter

251

How does the thermal type gauge compare with earlier 'cross-coil' type gauges?

....

....

....

....

An alternative to the thermal (bi-metal) voltage stabiliser is the electronic type shown below.

Describe the operation of this voltage stabiliser.

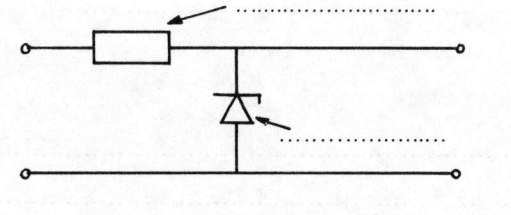

....

....

....

....

FLUID LEVEL SENSORS

Low fluid level warning for brake fluid reservoir, windscreen washer reservoir etc. can be sensed by the use of a float and magnet operating a 'reed' switch. Examine a level sensor switch and make a simple sketch to illustrate its action.

VEHICLE ....

LEVEL INDICATED

Low engine oil level can be indicated by the use of a HOTWIRE sensor located inside the 'dipper' end of the dipstick. Complete the sketch to show this type of sensor and explain how it works.

....

....

....

....

....

....

PRESSURE SENSORS

Oil pressure can be sensed by a pressure switch which will operate a warning light at a pre-determined low pressure.

Complete the sectioned view of the switch and explain how it operates.

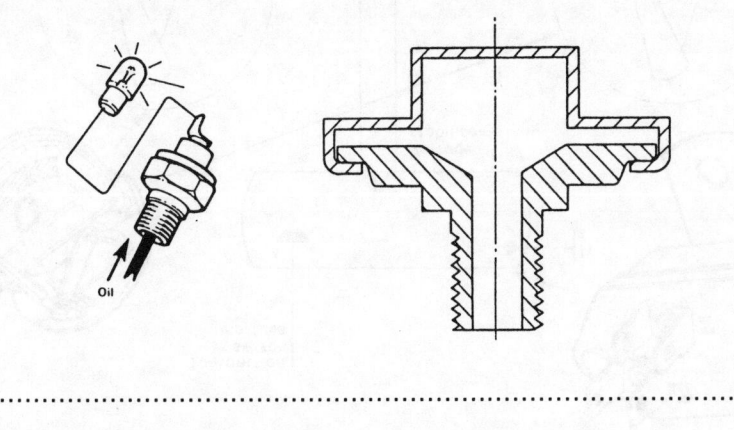

....

....

....

An indication of oil pressure can also be provided by a thermal type pressure sensor in conjunction with a thermal gauge.

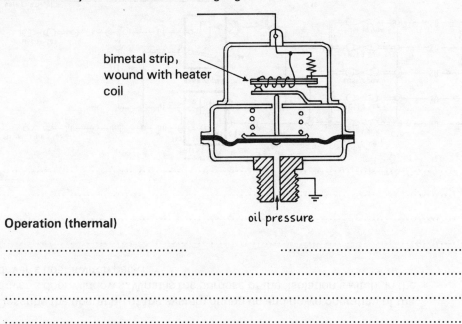

bimetal strip, wound with heater coil

oil pressure

Operation (thermal)

....

....

....

....

....

TURBO BOOST SENIOR

A pressure-sensing transducer can be incorporated into the turbo charging system to operate, in conjunction with ECU, wastegate, solenoid valves etc., to control boost pressure in relation to engine operating requirements.

Sketch and describe one type of pressure sensor which could be used for this purpose:

....

....

....

....

....

....

....

SPEEDOMETER/TACHOGRAPH

The speedometer/tachograph can be driven by an electric motor at the rear of the instrument. A transducer, which is a quenched oscillator pulse generator driven from the gearbox, provides current pulses which are fed to the instrument ECU. The speed of the electric motor is regulated by the ECU in response to the signal from the speed transducer. Describe, with the aid of a simple diagram, the operation of the pulse generator.

ELECTRICALLY OPERATED WINDOW

Side windows can be operated by electric motors connected to a geared linkage which moves the windows up or down, depending on the motor direction of rotation. A dc permanent magnet motor is used to power each window drive mechanism.

How is the direction of the motor reversed for up or down window operation?

....

....

....

....

The circuit below shows the main front/rear relays which are energised (fuse 16, wire GY) when the ignition is switched on.

Describe the operation of the system during up and down operation of the driver's door windows. What is the purpose of the 'isolation switch' in the driver's door switch pack?

....

....

....

....

....

Window operation

Up

....

....

....

....

....

Down

....

....

....

....

....

Isolation switch

....

....

....

....

53.5 What is the difference between semi-automatic and fully automatic window switching systems?

....

....

....

....

Examine an electric window mechanism and make a simple sketch below to illustrate the motor drive mechanism.

CENTRAL DOOR LOCKING

A central door locking system enables all doors, boot or tailgate to be locked or unlocked simultaneously when the key is turned in the door lock. Electric motors or solenoids in each door and boot or tailgate are activated to operate the door locking mechanism. A front door lock and control is shown below; label the drawing.

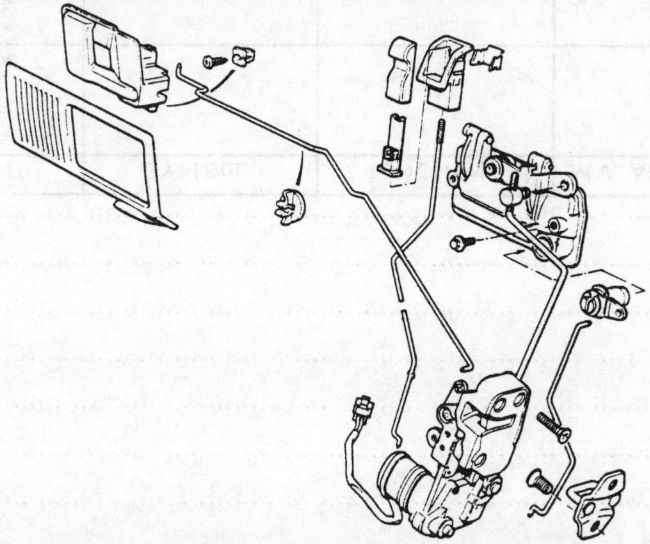

ELECTRIC WINDOW/CDL FAULTS, SYMPTOMS, PROBABLE CAUSES AND CORRECTIVE ACTION

Complete the table in respect of the faults listed.

FAULTS	SYMPTOMS	PROBABLE CAUSES	CORRECTIVE ACTION
Partial seizure of mechanism			
Circuit fault			

255

A complete central door locking electrical circuit is shown below. The ECU provides a current pulse to activate the locks. Study the circuit and describe the operation of the system during door lock and unlock.

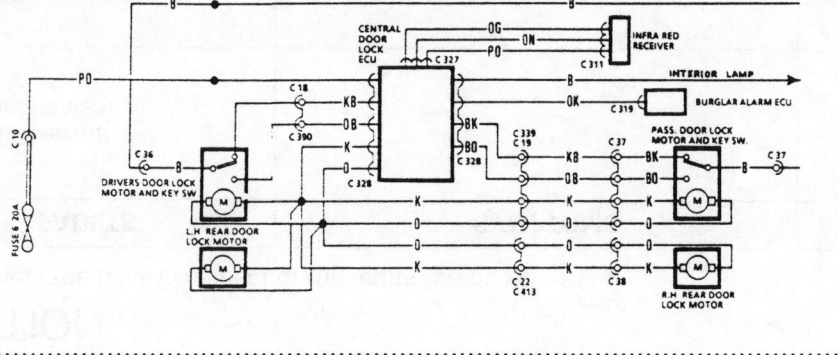

....
....
....
....
....
....

ELECTRONIC COMPONENTS
Complete the table.

COMPONENT	SYMBOL	FUNCTION	M.V. APPLICATION	COMPONENT	SYMBOL	FUNCTION	M.V. APPLICATION
transistor				thermistor			
Zener diode				thyristor			
L.E.D.				diode			

INFRA-RED LOCKING/UNLOCKING

The central door locking circuit can be energised using the door key or remotely using a hand-held infra-red transmitter.
The main components in an infra-red system are:
transmitter, receiver/sensor units, and ECU (see CDL circuit).
The transmitter consists of a coded integrated circuit, batteries, switch, emitter diode and function diode.

Note: Some remote control units use a miniature radio transmitter/receiver rather than infra-red, in order to activate the system.

State where the receiver/sensor units are located and describe the operation of the system.

....
....
....
....
....
....

IN-CAR ENTERTAINMENT

The radio and radio/cassette player have become almost universal fitments in today's cars. The range of equipment available for in-car entertainment, and the industry involved in it, is considerable. A relatively recent audio item is the COMPACT DISC (CD) player. Describe this:

CD PLAYER

....
....
....
....

The sound quality of the audio equipment is obviously dependent on the quality of the equipment: radio, cassette player, speakers, amplifiers, aerial etc. The GRAPHIC EQUALISER is now a popular ancillary item which enhances the equipment. State the function of a graphic equaliser.

GRAPHIC EQUALISER

....
....
....
....

The reception and reproduction of sound by a system relies not only on the quality of the equipment but also on the quality of the installation. Most reception and sound deficiences are due to problems with the installation. Complete the drawing top right to show the installation wiring for a radio and list the important points to be aware of when installing a radio.

....
....
....
....
....
....

SPEAKERS

Single, two, four or more speakers may be used in an audio system. The use of four speakers with a 'fader' control will give a better overall sound than, say, a two-speaker system. It is important to ensure that the IMPEDANCE (ratio of voltage to current in an ac circuit) of the speaker system is matched to the amplifier.
Add the wiring to the speaker layout shown.

AERIAL/ANTENNA

A high percentage of service problems can be attributed to inferior or badly fitted aerials. The ideal aerial is a fixed-length antenna but the practicalities of a telescopic aerial make it more suitable (against vandals, while car washing etc.).

Name two other types of aerial:

....

....

Siting

State the factors to consider when choosing a place on the vehicle to position the aerial.

....

....

....

....

Describe the procedure for fitting a mast type aerial on to the car body and complete the drawing to show an installation.

....

....

....

....

....

....

....

....

body panel

What is meant by 'aerial trimming'?

....

....

....

ELECTRICALLY OPERATED AERIALS

A permanent magnet electric motor can be used to power the drive mechanism for extension or retraction of the aerial mast. A separate two-way switch or a switch incorporated into the radio will operate the motor in opposite directions of rotation. Add the circuit to the aerial assembly shown and describe the drive mechanism for the aerial.

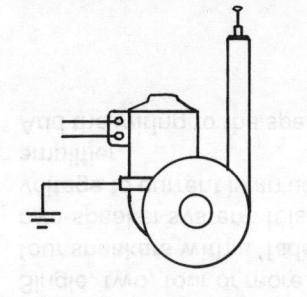

....

....

....

....

....

....

TWIN SYSTEM

Two antennas, one front wing, one rear screen can be connected via a computer to a receiver which has two tuners. How does this system operate and what are its advantages?

....

....

....

....

....

53.5 INTERFERENCE SUPPRESSION

Radio interference is caused by high-frequency waves being fed to the receiver, resulting in impaired reception. Interference can reach the receiver by both conduction and radiation.

Describe briefly how frequency interference can reach the receiver via each route.

Conduction interference:

....

....

....

....

....

....

Radiation interference:

....

....

....

....

....

High-frequency interference waves are produced by electrical components on a vehicle in which sparking or arcing occurs during operation. The main source of interference on a petrol-engined vehicle is the ignition system. List other sources of interference on a motor vehicle:

....

....

....

A further source of interference is the build-up of an electro-static charge in body panels, resulting in electrical discharges to adjacent panels. Methods of suppressing interference include the use of: resistors, capacitors, chokes, screens and straps.

How is the high tension ignition circuit suppressed?

....

....

....

....

Show, by sketching, the installation of a capacitor on the components below.

Give examples of the use of CHOKES in electrical circuits for suppression purposes:

....

....

SCREENING AND STRAPS

Screening means creating a barrier between the source of interference and the aerial. Interference waves generated inside an effective screen cannot be radiated through it. Metal body parts such as the bonnet have a screening effect.

Metallic screening of the ignition system is an effective but expensive form of suppression.

What does this involve?

....

....

....

....

Earthing straps are an essential part of the interference suppression system. State the reason for these and give examples of places on a vehicle where they would be used.

....

....

....

....

....

Certain types of bodywork present particular screening problems. Why is this?

....

....

State the legal requirements relating to suppression on vehicles:

....

....

....

FAULTS (IN-CAR ENTERTAINMENT)

State the possible causes of the problems listed below and describe corrective action to overcome them.

FAULT	CAUSE	ACTION
Poor reception		
None or intermittent operation		
Loss of memory		
Physical damage		

Suggest ways in which in-car-entertainment systems can be upgraded.

....

....

....

....

....

ANTI-THEFT SYSTEMS

Vehicle anti-theft systems range from a simple immobiliser switch in the ignition circuit, to a highly sophisticated electronic alarm system.

Alarm systems differ basically in the method of sensing interference with a vehicle and in the type of alarm warning activated (such as vehicle horn, siren, flashing lights personal bleeper etc.).

Describe briefly the following types of alarm system:

Voltage-drop type

....

....

....

....

....

Inertia type

....

....

....

....

Ultrasonic type

....

....

....

....

Examine one type of alarm system and make a sketch below to show the circuit. Include: switches, sensors, control and warning device.

Describe any adjustments which may have to be made to a system following installation

....

....

....

An alarm system can be extended to provide specific protection for items such as spot lamps, radio cassette, caravan. This entails the use of a '24 hour loop' system. Describe this:

....

....

....

....

53.7 To measure and assess the condition of electrical circuitry and components, certain checks and performance tests are necessary. Describe the checks and tests associated with the following.

Switches/relays

....
....
....
....

Sensors/voltage stabilisers

....
....
....
....

Wipers and washers

....
....
....
....
....

Horns

....
....
....
....

Instrumentation

....
....
....
....

Electric window lift

....
....
....
....

Central door locking

....
....
....
....

Anti-theft devices

....
....
....
....

How might electrical systems be protected against hazards during use or repair?

1. ....
2. ....
3. ....
4. ....
5. ....
6. ....

List general rules for efficiency and any special precautions to be observed when testing, overhauling and repairing the electrical system:

1. ....
2. ....
3. ....
4. ....
5. ....
6. ....
7. ....
8. ....
9. ....

TESTING ELECTRONIC COMPONENTS

Describe, with the aid of simple sketches, how to test the components on this page.

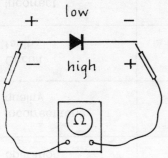

The ohmmeter should show a low reading when connected in one direction and a high reading when connected the opposite way round.

DIODES

THERMISTORS

TRANSISTORS

THYRISTORS

263

53.8 AUXILIARY ELECTRICAL FAULTS, SYMPTOMS, PROBABLE CAUSES AND CORRECTIVE ACTION

Complete the table in respect of the faults listed.

	FAULTS	SYMPTOMS	PROBABLE CAUSES	CORRECTIVE ACTION
WINDSCREEN WIPER/WASH	Play in linkage			
	Worn or partial seizure of drive gear			
	Weak wiper arm springs			
	Worn wiper blade rubbers			
	Faulty washer operation			
HORN	Incorrect tone quality			
	Circuit fault			
	Incorrect gauge reading			
	Incorrect alarm system operation			

Chapter 14

Related Studies

ELEMENT 60

ELECTRON THEORY

The 'Electron Theory' gives a basis by which the flow of electricity can be visualised and understood. Enlarge upon the following points in the Electron Theory:

1. Electricity in a conductor is in the form of electrons.

2. The flow of electric current is a movement of electrons.

3. The electrons may be forced to move by suitably applied magnetism or chemical pressure.

....

....

....

....

....

....

Make simple sketches to show the negatively charged electrons, positively charged protons and neutrons of a copper atom, showing why it is a good conductor of electricity.

In terms of electron flow:

a conductor is a material that will ....

....

....

an insulator is a material that will ....

....

....

a resistor is a material that will ....

....

....

The connections in an electrical circuit are said to be either positive or negative.

The flow of an electric current is said to pass

from .... to ....

and the flow of electrons is said to pass

from .... to ....

....

....

State what the following electrical terms mean, and give the units in which they are measured.

Voltage ....

....

Current ....

....

Resistance ....

....

PRODUCTION OF ELECTRICITY

Electricity can be produced:
chemically, magnetically or thermally.

Chemical

Label the simple cell shown.
Copper and zinc plates in
dilute sulphuric acid constitute
a simple cell.
The potential difference between

SIMPLE CELL

the positively charged copper plate and the negatively charged zinc plate is
about 1 volt (electromotive force – emf). If a lamp is connected as shown, the
emf will cause a current to flow:
Describe briefly with the aid of simple sketches production of electrical potential
by:

Magnetic means:

....

....

....

....

....

....

Thermal means:

....

....

....

....

....

EFFECTS OF AN ELECTRIC CURRENT

The flow of current in an electrical circuit produces three main effects. These
are: HEATING, CHEMICAL and MAGNETIC.
Describe the three effects stated above.

Heating

....

....

....

....

....

Chemical

....

....

....

....

....

....

Magnetic

....

....

....

....

....

Give motor vehicle applications of the three electrical effects.

Heating ....

Magnetic ....

Chemical ....

....

OHM'S LAW

'Ohm's Law', is the expression that relates voltage, current and resistance to one another.

Ohm's Law states that:

....

....

....

....

Ohm's Law as a formula using symbols can be expressed as:

$$I = \text{...}$$

$$I = \frac{V}{R} \quad \text{where} \quad V = \text{...}$$

$$R = \text{...}$$

This simple equation can be used to calculate any one of the three values provided the other two are known.

Resistivity

The unit of resistivity is the OHM – METRE and the resistance of a conductor is dependent on:

1. Material (resistivity).

2. ....

3. ....

4. ....

Resistance (Ω) = $\dfrac{\text{\hspace{2cm}}\times\text{\hspace{2cm}}}{\text{\hspace{4cm}}}$

State the resistivity of copper: ....

At a constant temperature, the resistance of a wire will vary proportionally to the dimensions of that wire (that is, the length and cross-sectional area).

How is the resistance of most metals affected by temperature change?

....

Problems

1. What voltage will be required to cause a current flow of 3 A through a bulb having filament resistance of 4.2 ohms?

2. What will be the total resistance offered by a lighting circuit if a current of 11 A flows under a pressure of 13 V?

3. Two 12 V headlamp bulbs each have a resistance of 2.4 ohms. Calculate the current flowing in each bulb and the total current flowing in the circuit.

4. A cable has a resistance of 0.4 Ω and an area of 15 mm². If the area is increased to 90 mm² what would be the resistance?

5. If the resistance of a wire is 0.009 Ω when it is 6 m long, what will be its resistance when it is 72 m long?

6. A copper cable has a CSA of 105 mm. Calculate the resistance of a 30 m length of the cable. (Resistivity of copper = 1.72×10^{-8}).

ELECTRICAL CIRCUITS

To allow an electrical current to flow an electric circuit must consist of:

(a) A source of supply.
(b) A device that will use the supply to do useful work.
(c) Electrical conducting materials that will transfer the electric current from the supply source to the consuming device, and then return it to the supply source.

On a motor vehicle TWO sources of electrical supply are:

1. ....

2. ....

Name FIVE different types of devices that consume the current to do useful work:

1. ....

2. ....

3. ....

4. ....

5. ....

What is meant by the term electrical conductor?

....

....

What is meant by the term electrical insulator?

....

....

Name SIX electrical conductors and SIX insulators:

Conductors	Insulators
1.	1. ....
2.	2. ....
3.	3. ....
4.	4. ....
5.	5. ....
6.	6. ....

INVESTIGATION

Connect the following components to build an electrical circuit:

battery, ammeter, switch, light bulb.

Using conventional symbols draw the circuit diagram. Indicate using arrows the conventionally accepted direction of current flow.

State the amount of current flowing in the circuit shown ....

The bulb wattage was ....

MEASURING INSTRUMENTS

An .... measures the amount of current flow. This instrument when used, must always be connected into the circuit in ....

When checking the voltage (or potential difference) across components in a circuit a .... is used. This must be connected across the terminals of the components being tested, that is in ....

To check the resistance of an electrical component, for example coil, the meter used is called an .... In what way does this meter differ from the other two? ....

....

Check the resistance of the following:

coil ...., field winding ...., small light bulbs ....

BUILDING VARIOUS ELECTRICAL CIRCUITS

INVESTIGATION

Equipment:

　　various types of resistances; switches; cables; ammeter; voltmeter; battery

Assemble the circuits described below, include in each circuit an ammeter, switch and battery.

You may be provided with **either:**

　　(a) motor-vehicle components　　　(b) a peg board　　　(c) a construction kit

Sketch the circuit diagrams and state the total current flow in each case.

In a series circuit the bulbs or resistances are connected:

....

....

....

....

In a parallel circuit the bulbs or resistances are connected:

....

....

....

---

1. Two resistances connected in parallel

Current flow ....

2. Four resistances connected in series

Current flow ....

3. Six resistances connected in parallel

Current flow ....

---

4. Two resistances in series connected with one resistance in parallel

Current flow ....

5. Two resistances in parallel connected to one resistance in series. Repeat this layout using three *additional* resistances and connect *both layouts* in parallel with one another.

Current flow ....

PARALLEL CIRCUITS

With the aid of diagrams state the basic laws of parallel circuits.

Voltage ....

....

....

Current ....

....

....

Resistance ....

....

....

Problems

1. Three conductors are placed in a parallel circuit, their resistances being 2, 3 and 4 Ω.
   What current will flow in each when connected to a 12 V system?

2. Two resistances of 20 and 5 Ω are connected in a 12 V parallel circuit.
   Calculate the total current flow.

3. Two resistances of 8 and 6 Ω are connected in parallel.
   What voltage would be required to cause a current flow of 7 A?

4. Four resistors of 6, 8, 10 and 12 Ω are connected in parallel to a 12 V circuit.
   Calculate the current flowing and the total resistance of the circuit.

5. Three resistors of 3, 5 and 6 Ω are connected to a 12 V battery.
   Calculate the total circuit resistance.

SERIES CIRCUITS

With the aid of diagrams state the basic laws of series circuits.

Voltage ....

....

....

Current ....

....

....

Resistance ....

....

....

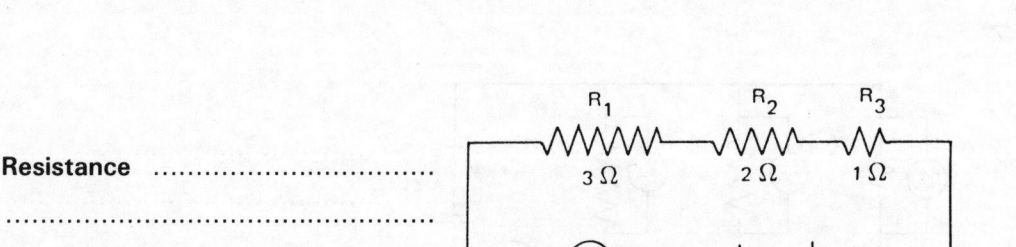

Problems

1. Two resistances of 1.75 Ω and 4.25 Ω are connected in series. What voltage would be required to cause a current of 2.5 A to flow in the circuit?

2. Four resistors of 6, 8, 10 and 12 Ω are connected in series to a 12 V circuit.
    Calculate the total resistance of the circuit and the current flowing in each resistance.

3. Four resistors of equal value are placed in series and connected to a 110 V supply. A current of 5 A then flows.
    Calculate the value of each resistor and the voltage across each resistor.

4. Three resistors are wired in series and when connected to a 12 V supply a current of 6 A flows in the circuit.
    If two of the resistors have values of 0.5 and 0.8 Ω calculate the value of the third resistor.

5. Three resistors of 2, 4 and 6 Ω are connected in series to a 12 V battery.
    Calculate the current flowing in the circuit and the voltage across each resistance.

PERMANENT AND ELECTRO-MAGNETS

There are two forms of magnets, permanent and electro-magnets.

What is the difference between a permanent and an electro-magnet?

....

....

....

These two forms of magnetism lead to the important relationship between magnetism and electricity:

....

....

INVESTIGATION

To produce an electro-magnet:
Equipment
    Coil of insulated wire and iron bar
    suitable for passing through centre
    of coil
    Resistance
    Battery
    Screwdriver or bar for checking
    magnetism.

Show sketch of apparatus used.

Tests

1. Test bar for magnetism
2. Pass current through coil
3. Test bar for magnetism
4. Attempt to pull bar out of coil. Switch off current.
5. Test bar for magnetism.

Effects of test were:

1. ....
2. ....
3. ....
4. ....
5. ....

What forms the basic construction of an electro-magnet?

....

The core of the electro-magnet may be moving as shown or stationary (for example, the ignition coil). In either case it is made from soft iton. Why is such a material used?

....

....

....

List FOUR motor vehicle components where an electro-magnet which produces movement is used.

1. ....
2. ....
3. ....
4. ....

LINES OF FORCE

Magnets act through lines of force. These lines of force stretch between the ends of a magnet and create a magnetic field. The two ends of a magnet are called .... one end being the ....

and the other the ....

Using small bar magnets, a sheet of paper and iron filings, show the effects of the lines of force when the magnets are held in the positions shown.

A [n s]   B [n s] [n s]   C [s n] [n s]

The effects created by the magnets lead to the statements:

Like magnetic poles .... Shown by sketch ....
Unlike magnetic poles .... Shown by sketch ....

273

ENERGY AND POWER

Electricity is a form of energy:

ENERGY (J) = VOLTS × AMPS × TIME (s)

or   Q   =

Power is the rate of doing work (energy per unit time):

Power (J/s)   (1 J/s = 1 Watt)

Electric power is measured in ....

Efficiency (η) is expressed as a percentage.

Efficiency in terms of energy and power is:

$$\frac{\text{Energy output}}{\text{Energy input}} \times 100 \text{ or } \underline{\hspace{3cm}} 100$$

Calculate the total energy consumed when a current of 3 amps flows in a 12 V ignition coil for 5 seconds.

A starter motor is operated for 30 seconds and the current flow is 200 amps. Calculate the energy consumed and the input power if the terminal voltage is 10 volts.

The power output of a starter motor is 1800 W. If the terminal voltage is 10 V and the current flow during starter operation is 240 amps, calculate the efficiency of the starter motor.

MUTUAL INDUCTION

A voltage can be INDUCED into a circuit by varying the current flow in an adjacent but separate circuit. This property is called MUTUAL INDUCTION. Label the drawing and describe how the simple transformer shown can STEP-UP input voltage.

....
....
....
....
....
....
....

How does the ratio of turns on primary and secondary windings affect the 'step up' voltage?

....
....
....

State one motor vehicle application of this principle:

....

MOTION

(a) ROLLER OR GEAR
    OPERATION

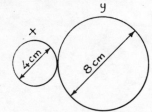

(b) BELT OR CHAIN
    OPERATION

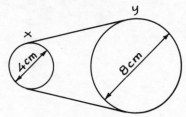

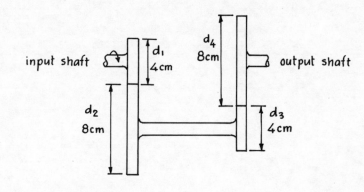

The method of transmission shown above is ....

Indicate the direction of rotation of the larger wheel in each case.

Rotational speed is expressed in ....

Circumferential speed is expressed in ....

Calculate the rotational speed of wheel y when the speed of wheel x is 50 revs/s.

Let: d_1 = dia. wheel x d_2 = dia. wheel y

      n_1 = speed wheel x n_2 = speed wheel y

    Rotational speed wheel y =

    ――――――――――――― =

Circumferential speed = $\pi d n$

Circumferential speed wheel x =

Circumferential speed wheel y =

    $\therefore$ $d_1 n_1$ = $d_2 n_2$

What does this prove about the relationship between the wheels in drive trains such as those shown above?

....

....

....

The two pairs of gearwheels shown above provide a gear reduction between input and output shafts. This method of transmission (as employed in a motor vehicle gearbox) is .... It is a COMPOUND gear train.

How does the direction of rotation of the input shaft compare with that of the output shaft?

....

The speed of the output shaft relative to that of the input shaft depends on the overall gear ratio. To determine overall ratio, the ratio of the first pair of gears is multiplied by ....

....

Use the diameters of the gears shown above to calculate overall ratio and output shaft speed when input speed is 1000 revs/s.

By what other methods can the ratio be obtained?

....

Chapter 15

Industrial Studies

UNIT 22

COMMERCIAL ASPECTS

The vehicle retail and repair industry is an essential part of the life of the UK. Like other major industries, it provides a contribution to the wealth of the nation or the 'national economy'. That contribution to the national economy is substantial and is made in a variety of ways.

Give four examples of how the industry contributes to the national economy of the country.

(a) Provision of jobs

....

....

....

....

....

....

....

State the simple commercial concepts governing the operations of the vehicle retail and repair industry, in the following respects:

(a) Supply and demand in the context of products and services

....

....

....

....

....

....

....

(b) Influence of capital costs, overheads, materials, labour and inflation on pricing and profit

....

....

....

....

....

....

....

....

(c) The essential aims of providing the right product at the right price, quality, timescale and after-service.

....

....

....

....

....

....

....

THE ROAD TRANSPORT INDUSTRY — SECTIONS

When considering the development of the Road Transport Industry from its early formative years around the beginning of this century, it is useful to identify the major sections of the industry. The industry is concerned with:

Vehicle and component manufacture ....

....

....

....

....

....

PRODUCTS

Manufacture

The divisions within the manufacturing side of the industry are:

Vehicle manufacture ....

....

....

....

While the vehicles are designed and built by the vehicle manufacturers, many of the components and systems are designed and manufactured by specialist companies. Give typical examples of components and systems which are supplied to vehicle manufacturers from specialist firms.

....

Ignition and electrical components and systems (Lucas) ....

....

....

....

....

....

The fuels and lubricants industry is of vital importance to road transport. List below some of their most important products.

Leaded and unleaded petrol ....

....

....

....

....

The fitting of vehicle accessories is very big business. Make a list of those you consider especially desirable.

....

....

....

....

....

STRUCTURE

Goods being distributed from a manufacturer or producer to the consumer usually flow through a distribution network which is as follows:

MANUFACTURER → .... → .... → .... CONSUMER

In the motor industry vehicles, parts and accessories reach the consumer via the following channels.

VEHICLE/PARTS MANUFACTURER (PRODUCER)
↓
DISTRIBUTOR (.................................)
↓
DEALER (.................................)
↓
CONSUMER (.................................)

Name two distributors and two of their respective dealers in your area; state the manufacturer in each case.

1. Manufacturer .... 2. ....

Distributor ....

Dealer ....

In the motor trade there is a move to delete one tier of the sales distribution network. Thus car sales are likely to be

Manufacturer   Dealer   Customer

State two likely consequences of this move:

....

....

....

List the main services provided by the sales and servicing side of the motor industry.

New and used vehicle sales

....

....

....

....

....

....

....

....

....

....

....

The main departments in a typical dealership are

New and used vehicle sales

....

....

....

....

....

....

STAFFING

A garage workshop may employ TECHNICIANS, CRAFTSMEN and OPERATIVES. Describe briefly the role of each type of worker.

TECHNICIAN

....
....
....
....
....
....

CRAFTSMAN

....
....
....
....
....
....

OPERATIVE

....
....
....
....
....

WORKSHOP FOREMAN
His role is to allocate the jobs, liaise with the service manager and reception, supervise and inspect work in progress, ensure safe working practices are adopted and generally ensure that management policies are implemented.

State briefly the roles of the garage personnel listed below:

SERVICE MANAGER ....
....
....
....
....
....

RECEPTIONIST ....
....
....
....
....

CAR SALES STAFF ....
....
....
....

PARTS DEPARTMENT STAFF ....
....
....
....

ADMINISTRATIVE STAFF ....
....
....
....

ROAD TRANSPORT INDUSTRY ORGANISATIONS AND ASSOCIATIONS

The initials of certain prominent trade associations are listed below. State the full title of the association in each case and give brief details of their (similar) roles.

MAA ....

SMTA ....

These are associations of garage employers which represent them on various committees such as wage negotiations.
Employers can use the association for legal or other advice.

SMM&T ....

....

....

....

....

....

....

RHA and FTA ....

....

....

....

....

Name two trade unions which represent workers in the road transport industry.

....

How are 'national agreements' on pay, job classification, working conditions and dispute procedures worked out for the motor industry?

....

....

....

....

....

....

....

TRAINING ORGANISATIONS

The *1964 Industrial Training Act* sought to improve the QUALITY and QUANTITY of industrial training and to share the cost of training more equally among employers. To do this, statutory Training Organisations were set up. Name training organisations relevant to the road transport industry.

....

....

....

....

....

....

30.7 CUSTOMER RELATIONS

Establishing a favourable relationship with customers (goodwill) is vital if a business is to survive. It can take years to build up the goodwill of a business; it can, however, be destroyed very quickly. A customer's impression or perception of a firm is influenced by a number of factors. One important factor is in his or her dealings with the firm's employees. Give some examples of the things which may influence a customer one way or the other when dealing with employees.

(a) *The visual appearance of an employee* ....

....

....

....

....

....

....

....

....

The way in which customer complaints are handled can either boost or be damaging to customer relations.
List some of the important points to be aware of when dealing with a complaint.

(a) *Adopt a positive attitude* ....

....

....

....

....

....

....

....

FAIR TRADING

Under the terms of the *Fair Trading Act*, it is illegal to adopt trade practices which adversely affect the economic interest of the consumers. To enforce this, the local Trading Standards Officer can enter motor-trade premises, make test purchases and inspect goods etc.
To what does 'Trade Practice' relate?

(a) *Advertising, labelling and packaging of goods* ....

....

....

....

VOLUNTARY CODES OF PRACTICE

One of the four voluntary codes of practice for the motor trade is published jointly by the MAA, SMTA and the SMM&T. What is a voluntary code of practice on fair trading?

....

....

....

....

....

....

CAREERS IN ROAD TRANSPORT

The career pattern for a person employed in the road transport industry is really determined by:

(a) The particular branch or area of road transport he or she is employed in, for example, car retail and repair, freight transport, PSV operation.

(b) The policies adopted by the employee's company — for example, a transport manager may be an ex-HGV driver or an ex-workshop fitter; a garage receptionist may be an ex-motor vehicle technician or a person with little or no knowledge of motor repair.

A typical career pattern for an apprentice motor vehicle mechanic engaged in freight transport might be

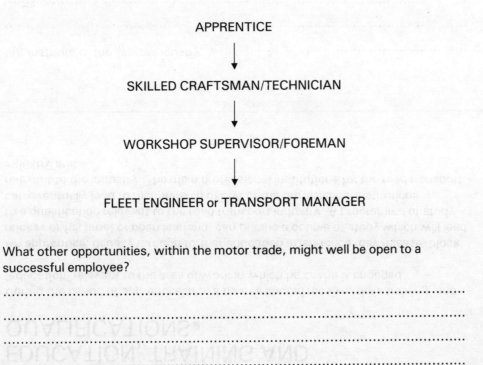

APPRENTICE

↓

SKILLED CRAFTSMAN/TECHNICIAN

↓

WORKSHOP SUPERVISOR/FOREMAN

↓

FLEET ENGINEER or TRANSPORT MANAGER

What other opportunities, within the motor trade, might well be open to a successful employee?

....

....

....

Choose any one of the following work areas and produce a flowchart to show a typical pattern for the career chosen.

WORK AREAS

Vehicle repair, vehicle sales, parts department, PSV operation.

....

....

....

....

....

....

Suggest personal qualities that you consider important to further a successful career pattern in the work area you have chosen above.

....

....

....

....

What do you see as your own hoped-for career pattern?

....

....

....

....

....

What, realistically, do you need to do in order to achieve your objectives?

....

....

....

EDUCATION, TRAINING AND QUALIFICATIONS

During the period of apprenticeship, a trainee should receive 'skills training' and 'education' relevant to the area of work in which he or she is engaged.

An apprentice, usually through further education attendance (day release, block release or full time) or open learning, can pursue a course of study which will lead to a qualification relevant to the road transport industry. A programme of study can eventually lead to membership of one of the professional institutions relevant to the industry. The main professional institutions for the road transport industry are:

(a) *Institute of the Motor Industry* ....

....

....

....

On the motor vehicle repair side of the industry, courses of study and qualifications are drawn up by the various examining bodies and training organisations, such as City and Guilds, Royal Society of Arts, the National Joint Council or the Road Transport Industry Training Board.

Produce a flowchart below to show the course programme and qualifications for an apprentice motor mechanic.